The Lightning Rod as a Danger

Jan Meppelink

The Lightning Rod as a Danger

Springer

Jan Meppelink
High Voltage Engineering
Emeritus University of Applied Sciences
Soest, Germany

ISBN 978-3-031-30433-0 ISBN 978-3-031-30434-7 (eBook)
https://doi.org/10.1007/978-3-031-30434-7

This Springer imprint is published by the registered company Springer Nature Switzerland AG
The registered company address is: Gewerbestrasse 11, 6330 Cham, Switzerland

With knowledge, doubt grows.

Johann Wolfgang von Goethe

Preface

LPS designers need to be clear about the effects and hazards to people and animals from a lightning rod. These are essentially the step voltage, the touch voltage, the induced voltage and the danger of a side flash when a person approaches a lightning rod. These effects can also occur in combination. These hazards must be limited to non-hazardous values by appropriate measures, so that injuries and death are safely avoided.

The IEC 62305 standard only provides general information on this subject. For a more precise planning of measures or for the evaluation of existing lightning protection systems, knowledge of the permissible limit values for step voltage and touch voltage is required. Due to some recent publications, a value of 25 kV of form 10/350 has been included in ED 3 of IEC 62305-3 as an international compromise for the step voltage.

In the literature, a touch voltage of 2 kV is specified as the permissible touch voltage for the lightning current 10/350. This is only valid for galvanic coupling. In the case of inductive coupling, the specification of the voltage does not make sense, since the voltage form deviates strongly from the 10/350 form. In these cases, the values W and Q must be calculated. Limiting values valid according to the state of the art are named in [3] and are $W_G = 1$ Ws for the energy converted in the body with an equivalent resistance of 1 kOhm and $Q_G = 1$ mAs for the charge. Therefore, an international agreed value for touch voltage is under discussion and may be published in the next ED 4 of IEC 62305-3 about 2030.

For a decision as to whether measures are necessary or not, an evaluation of the hazard potential is required. This can be done by calculations. In recent years, numerical field calculations have become increasingly user-friendly and are now so powerful that 3-D simulations can be performed on home computers in an acceptable time frame. The limit values can be used directly with the help of the routines stored in the programmes or can be easily checked by post-processing.

With the possibility of calculating even complex systems, the questions of the experts from the specialist circles also became more demanding. For example, questions arose about the effect of site insulation and its behaviour at high electric field strength and the possibility of the formation of surface discharges. Other questions include the effect of water-covered site insulation. Newer materials, such as water-permeable pavements are in use. Insulating down conductors have been in use for a decade to reduce separation distance. Insulating down conductors have also recently been discussed for protection against contact and are the subject of a new Standard IEC 62561-9 which is under development in IEC TC 81.

However, in the course of discussions in the technical circles and with experts, questions arose as to the necessity of such measures in relation to the injuries and fatalities actually occurring in the world due to step voltage, touch voltage and proximity. Based on the literature, it appears that only a small portion of about 5% of all injuries and fatalities are due to touch voltage. The main cause, however, is step voltage at 50% and side flash at 30%. The way people behave during a spontaneously occurring thunderstorm plays a major role. It is not likely that a person touches a down conductor just at the moment of the lightning strike. However, it is more likely that a person in the open field is surprised by a lightning discharge and is affected by step voltage.

The evaluation of the risk component R_A for injury to human beings (injuries, including loss of life, to people resulting from lightning) by electrical shock due to step and touch voltage according to IEC 62305-2 shows very low values. The meantime between two dangerous events with death or injury calculated for a normal one family house without lightning protection system shows a value of 13 million years. According to IEC 62305-2 the representative tolerable risk is 10,000 years. This leads to the conclusion that a lightning protection system is not required for a single-family house.

This Technical Report is intended to provide answers to these questions.

Suggestions for continuous improvement and expansion are welcome and will be considered in the second edition of this report.

Soest, Germany Jan Meppelink

Acknowledgements

The author thanks the "Committee for Lightning Protection and Lightning Research of VDE (ABB)" for its support for the study of step voltages.

Special thanks to the support engineers of Comsol-Multiphysics and XGS-Lab in solving special simulation problems.

Special thanks to Prof. Dr. Volker Hinrichsen for making the high voltage laboratory of the TH-Darmstadt available for the measurement of voltage-time curves.

Special thanks to my students who performed many measurements, especially to Cornelius Epple for his support in laboratory tests and solving problems with simulations in Comsol-Multiphysics.

This report has been translated into English with the help of DeepL.

About This Book

This report investigates the physical effects of a lightning flash on a person near the down conductor of a lightning protection system. These effects are the touch voltage, the step voltage and the side flash. For this purpose, a full-scale simulation model of the human body with a resistance of 1000 ohms was first created. In the simulation model, the body can touch the down conductor or be placed close to it. Furthermore, the specific resistance of the earth is varied. Likewise, insulating layers such as asphalt can be incorporated into the simulation model. Also, special cases like water-permeable layers or water layers on an asphalt layer can be calculated. In post-processing, all relevant values can be determined, such as the energy converted in the body, the charge, the current and the voltage applied to the body. A comparison with the permissible limit values shows for the lightning protection classes whether there is danger or not and provides information on necessary measures.

There is a risk for death and injury if the down conductor is touched. However, there is also a risk of a side flash if a person is standing next to a discharge. Site isolation with dry asphalt is effective, but there is a residual risk of surface discharges. In real situations with wet asphalt, water-permeable layers or asphalt with a water layer, however, there is a great risk of death or injury. Equipotential bonding with an earthing grid is a necessary but not sufficient solution with regard to the induced voltage at negative subsequent stroke. Therefore, the situation must always be examined on a case-by-case basis with regard to the safety requirements. The only effective measure to prevent injury and death due to touch voltage is an insulating down conductor in conjunction with equipotential bonding.

The measures for reducing the touch voltage, such as site insulation and equipotential bonding, basically also apply to limiting the step voltage.

A risk calculation according to IEC 62305-2 gives the mean time between two events of injury and death $M_G = 1/R_A$. The tolerable risk is: $R_A = 0{,}0001$ or $M_T = 10.000$ years, equivalent to one death in 10.000 years. An example for a building with dimensions 15 m x 15 m x 10 m shows the following values. These figures fit the trend of actual published values of death and injury from touch voltage. However, these numbers do not fit at all with the published values for death and injury from step voltage. IEC 62305-2 does not specify the risk of death or injury from side flash.

No lightning protection system present:	With lightning protection system class III/IV
Arable soil M_G = 13,7 million years	Arable soil M_G = 137 million years
Gravel: M_G = 137 million years	Gravel: M_G = 13,7 billion years
Asphalt: M_G = 13,74 billion years	Asphalt: MG = 137,4 billion years

Instead of a comprehensive summary, the most important questions and answers are documented as an FAQ list as follows:

FAQ's

In the following, the most important questions associated with step voltage, touch voltage and the risk of a side flash are answered as an FAQ list. This is intended to simplify the reader's path to an answer by referring to the detailed descriptions, Sections and figures in the document.

Question	Normative references	Section
Can touch voltage cause death or injury? Yes, but it is very unlikely and depends on the probability that a person touches a down conductor just at the instant of a flash. Only 5 % of all reported injuries are related to direct contact; 15–20 % of all death cases are related to direct contact	IEC 62305-2	14, 16, 11.1 Table 29
Can step voltage cause death or injury? Yes, it is more likely compared to touch voltage. The probability that a person is located in the area around a down conductor or in the field in the area of the striking point is higher. 50–55 % of all reported injuries are related to step voltage; 45 % of all death cases are related to step voltage	IEC 62305-2	14,16, 11.2 Table 30
Can a side flash cause death or injury? Yes, it is more likely compared to touch voltage. The probability that a person stands close to a down conductor is higher compared to direct contact. 20–35 % of all reported injuries are related to side flash; 20–30 % of all death cases are related to side flash	IEC TR 60479-4	14,16

(continued)

(continued)

Question	Normative references	Section
What causes the touch voltage? The touch voltage is created by the current flow in the soil and the associated voltage drop between the current carrying down conductor and the foot of the person. In addition, the magnetic field acts in the area between the person and the down conductor, which induces a voltage. The voltage drop along the down conductor is negligible. The touch voltage is not coupled to a fixed distance. It depends on the distance of the person to the arrester	IEC TR 60479-4	2
What causes the step voltage? The step voltage is created by the current flow in the soil and the associated voltage drop between both feet. The step voltage is coupled to a fixed distance of 1 m between both feet. But the step voltage decreases with increasing distance from the current carrying down conductor	IEC TR 60479-4	10
What causes the side flash? The side flash is created by the current flow in the soil and the associated voltage drop between the current carrying down conductor and a person near to the down conductor. The person does not touch the down conductor. This is known from deaths where a person had taken shelter near a tree	IEC TR 60479-4	6.4
Limits for touch- and step voltage in terms of voltage or in terms of Energy and charge? For the step voltage caused by a first stroke having a shape of 10/350 a limit voltage of 25 kV of the shape 10/350 has been found as representative limit. In this case a pure galvanic coupling is predominant For the touch voltage both galvanic and inductive coupling occur. The shape of the induced voltage is proportional the di/dt of the current. Therefore, it does not make sense to set a voltage value. Therefore, according to the state of science, a limit value for the energy converted in the person of 1 Ws and a converted charge of 1 As is recommended For the 10/350 pulse, a value of 2 kV is suggested in the literature	IEC TS 60479-1	4 [34]

(continued)

(continued)

Question	Normative references	Section
What is a representative model for the human body? A person can be simulated by a resistor of 1000 ohms. It is therefore possible to use a human body with its dimensions and adjust the conductivity of this body to create a resistance between hand and foot of 1000 ohms. Such a model can then be used with a numerical field calculation programme, e.g. Comsol-Multiphysics. This allows in post-processing the determination of the energy and charge converted in the body of a person	IEC TS 60479-1	2 [34]
What about ionisation in soil? Soil ionisation decreases the resistance to earth especially at large values of soil resistivity. Without considering soil ionisation, the calculation is always on the safe side and represents the worst case. Consideration of soil ionisation is therefore recommended in special cases with really high soil resistivity. The calculation time increases noticeably		[23]
What about the voltage drop along a down conductor? The voltage drop along a down conductor can be neglected compared to other components. Even the transient skin effect has no significant influence compared to other effects		6.1
Are the kc coefficients (IEC 62305-3) valid for all impulse currents? No! The kc coefficients were calculated exclusively for the negative subsequent stroke 0.25/100 and are presented in IEC 62305-3. They are used to calculate the separation distance according to IEC 62305-3. Therefore, for the lightning current forms 1/200 and 10/350 kc coefficients were recalculated for two earth electrode types A and B	IEC 62305-3	3

(continued)

(continued)

Question	Normative references	Section
Is the model applicable for the impulse shape 0,25/100 as per IEC 62305-1 or is a consideration of travelling waves necessary? A quasi-stationary calculation (no travelling waves considered) is permitted if the rise time of the impulse is about 10 times larger than the transit time in the system. Furthermore, there is the peculiarity that the wave propagation in Soil depends on the specific earth resistance. This condition must therefore be checked in each individual case In the present case with a vertical earth rod of 9 m length and a specific earth resistance of 250 Ohmm and a 0,25/100 impulse, the condition is only just fulfilled. In individual cases, this condition must always be checked On the other hand, a calculation including the wave propagation in the earth and along the down conductor above ground is a highly complex calculation		2
What soil resistivity value is representative? In this report, unless otherwise mentioned, the soil resistivity is set to 250 Ohmm as a representative average value	IEEE 80 2015	
Risk of death through touch voltage due to galvanic coupling? Here, the 10/350 is the greatest hazard. However, the calculations show that for all lightning current shapes, 10/350; 1/200 and 0.25/100, a person would die on contact with a down conductor in all lightning protection classes	IEC TR 60479-4	6.8 Table 9a
Risk of death through touch voltage due to inductive coupling? Here, the 1/200 and the 0.25/100 pulses are the greatest danger because of their greater current slope compared to 10/350	IEC TR 60479-4	6.8 Table 9b
Risk of death through touch voltage due to galvanic and inductive coupling? Galvanic and inductive coupling always occur together. Therefore, there is a risk of death on contact with a down conductor in all lightning protection classes	IEC TR 60479-4	6.8 Table 9c

(continued)

(continued)

Question	Normative references	Section
Can you calculate the risk R_A of death and injury from step and touch voltage? Yes, according to IEC 62305-2 the risk R_A con be calculated A calculation gives the mean time between two events of injury and death $M_G = 1/R_A$. The tolerable risk is: $R_A = 0{,}0001$ or $M_T = 10.000$ years, equivalent to one death in 10.000 years. An example for a building with dimensions 15 m x 15 m x 10 m shows the following values The following values M_G are obtained for different materials: **No lightning protection system present:** Arable soil $M_G = 13{,}7$ million years Gravel: $M_G = 137$ million years Asphalt: $M_G = 13{,}74$ billion years **With lightning protection system class III/IV** Arable soil $M_G = 137$ million years Gravel: $M_G = 13{,}7$ billion years Asphalt: $M_G = 137{,}4$ billion years These figures fit the trend of actual published values of death and injury from touch voltage in Sect. 14 However, these numbers do not fit at all with the values for death and injury from pacing stress in Sect. 14 IEC 62305-2 does not specify the risk of death or injury from side flash	IEC 62305-2	16 Table 30
Can potential control (equipotentialization) reduce the touch voltage? The galvanic coupling is less but not sufficient. There is still inductive coupling The person would die for all impulses in all LPS-classes, except for 0,25/100 Class II and III-IV		6.7 Table 8

(continued)

(continued)

Question	Normative references	Section
Does the danger of inductive coupling depend on the distance of a person to the down conductor? Yes, there will be a reduction of the mutual inductance but there is still a risk of death: For 0,25/100 in LPS-classes I and II down to 20 cm For 1/200 in LPS-classes I and II including 10 cm For combined 1/200 and 0,25/100 in all LPS-classes down to 10 cm		6.8.3 Fig. 49
What is the effect of site insulation with a layer of dry asphalt ρ = 10.000 Ohmm? **1. on the touch voltage?** A layer of asphalt reduces the galvanic coupling and thus also reduces the touch voltage. However, a detailed analysis of the values shows that, despite an asphalt layer, the limit values are exceeded and people would die when they come into contact with the down conductor. This applies to all lightning current forms and lightning protection classes **2. on the step voltage?** The step voltage is not critical **3. on formation of surface discharges?** Because of the high potential at the foot of the person, the formation of surface discharges occurs which are fatal for a person. **4. on formation of side flash?** If a person standing on asphalt approaches a down conductor, there is a danger of a side flash with fatal effect. This can be avoided when using an insulating down conductor	IEC 62305-3 IEEE 80 2015	7.1 Table 12 7.3 Table 13 Fig. 63

(continued)

(continued)

Question	Normative references	Section
What is the effect of site insulation with a layer of dry asphalt ρ = 10.000 Ohmm and an additional potential control (equipotentialization)? There is no danger to persons due to galvanic coupling. Only in lightning protection classes I and II are the limit values exceeded for the negative subsequent stroke 0.25/100, so there is a danger there A danger from side flash with inductive coupling only exists if the distance between the person and the down conductor is < 10 cm in lightning protection class I		7.5.5 17.2.2
What is the influence of a 2 cm high water layer on an asphalt layer? The consequences are fatal. A current flow exists from the down conductor through the person into the water layer and from the water layer through both legs of the person. The person would die in all lightning protection classes and with all lightning current forms due to touch- or step voltage		7.6 Fig. 73
What is the influence of a 2 cm high water layer on an asphalt layer and an additional potential control (equipotentialization)? The consequences are still fatal. A person would certainly die due to touch—or step voltage		7.7.1
Site insulation with gravel? Gravel is not suitable as insulating material because of the risk of flashover inside gravel. It may be useful in power frequency applications	IEEE 80 2015	8
What is a water-permeable site insulation? This is a sandwich of a water-permeable mixture of stone particles and plastic, a mineral mixture, a layer of crushed stones and a layer of clean minerals		9.1 Fig. 88

(continued)

(continued)

Question	Normative references	Section
What is the effect of site insulation with a water-permeable layer in wet conditions ? The consequences are fatal. A current flow exists from the down conductor through the person into the water layer and from the water layer through both legs of the person. The person would die in all lightning protection classes and with all lightning current forms due to touch- or step voltage		9.2 Table 24
What is the effect of site insulation with a water-permeable layer in wet conditions and an additional potential control (equipotentialization)? The consequences are still fatal. A person would certainly die due to touch—or step voltage		9.3 Table 27
What is the danger of step voltage? The step voltage depends on where a person stands in relation to the down conductor But the probability that his person is in a dangerous step voltage range is greater than the probability that a person touches the down conductor On the other hand, the limit value of 25 kV for step voltage valid for 10/350 is significantly higher than the value of 2 kV for touch voltage valid for 10/350 For a detailed discussion, see Sect. 21	IEC TR 60479-4 IEEE 80 2015	10 21
What type of earthing system is more dangerous, type A or type B? According to Sect. 10.3, it is clear that a type B grounding system presents a greater hazard. Therefore, all calculations in this report have been carried out for earth system A	IEC 62305-3	10.3

(continued)

(continued)

Question	Normative references	Section
What measures can be taken according to IEC 62305-3 to reduce touch and step voltage? **Distance:** It is recommended a distance of 3 m from a down conductor. According to the calculations in this report, it is an effective measure but difficult to achieve **Site insulation with a layer of 5 cm Asphalt** It applies only for dry conditions which are unrealistic in case of thunderstorms However, there is a risk of surface discharges. **Signage** Signage is not practical, as it would have to be executed in several languages and is not legible in the darkness **Insulation of the down conductor** IEC 62305-3 requires insulation of the down conductor with a layer of polyethylene at least 3 mm thick. A withstand impulse voltage of 100 kV 1.2/50 is required. It is not specified how this impulse withstand voltage is to be determined. The standard does not refer to the reduction of the electrical strength due to surface discharges. The specified value of 100 kV is unrealistic, as the previous calculations show. These discrepancies must also be eliminated in ED 4 of IEC 62305-3 A better solution to protect a person against touch voltage is the use of insulating down conductors. This is under consideration for a new work item proposal in IEC TC 81 as a future standard IEC 62561-9, *Requirements for components for protection against dangerous touch voltage*	IEC 62305-3	7.6
Can an isolating down conductor be a solution to reduce the touch voltage to harmless values? For economic reasons, only a combination of equipotential bonding and components for protection against dangerous touch voltage, e.g. insulating down conductors makes sense, and the remaining dangerous inductive coupling can be controlled	TS-IEC 62561-8	13 Table 25

(continued)

(continued)

Question	Normative references	Section
What is a surface discharge? This type of electrical discharge on the surface of an insulating material is listed in IEV Electropedia WWW.ELECTROPEDIA.ORG under IEV ref. 212-11-45 as: **surface partial discharge,** partial discharge along, or onto, the surface of an insulation. In this report, the term surface discharge is used. A surface discharge can be a partial discharge or a complete surface flashover In other references this discharge type is called sliding discharge or creepage discharge, however not in line with the IEV The physics of this very special type of an electrical discharge on surfaces of insulating materials is explained in detail in Sect. 21		21
Where can a surface discharge occur when looking at touch and step voltage? On those insulating arrangements where the electric field strength is perpendicular to the surface of the insulating material, e.g.: • On all insulated down conductors (Person touches the surface of insulation) • On insulated layers of asphalt or other highly insulating materials. (Person stays on the layer and touches the down conductor)		7,9

Contents

About the Author

Jan Meppelink received his Dipl.-Ing. Degree and Dr.-Ing. in electrical engineering from the Technical University of Berlin. 1984–1988 he was Leader of Basic Development High Voltage Engineering, ABB High Voltage Technologies, Zürich, Switzerland. He was Leader of the study group, Principle Development in High Voltage Engineering and Gas Insulated Switchgear (GIS), especially for the performance verification testing of 800 kV Gas Insulated Switchgear and commissioning of the 550 kV GIS Itaipu/Brazil, the development of the EMI proof design of secondary equipment for high voltage substations, research on propagation and measurement of very fast transients in GIS and development of ultra-fast voltage dividers for very fast transient measurements. 1998–1992 he was Manager Projects International Gas Insulated Switchgear, ABB High Voltage Technologies, Zürich, Switzerland. He was Leader of the Department for order handling and Project—Management, also Leader of the Engineering Department. Since 1992 Professor for High Voltage Engineering at the University of applied Sciences, Soest, Germany. Since 2008: Course Director of the Study Course, Engineering and Project Management. Field of Research: High voltage engineering, Lightning- and Over Voltage Protection and along with the study course engineering and project management, Lean Management, Process Management, Project Management, Sales Engineering. Since 2015 he is professor emeritus. He is Head of the German working group AK 251-04 dealing with standardisation of lightning protection system components, member of IEC TC81, MT8, MT14, MT21, and Cenelec TC81X WG2.

In 2021 he received the Benjamin Franklin medal from the German VDE/ABB.

Chapter 1
Introduction

In 1750, when Benjamin Franklin formulated the proposal to prove the effectiveness of a lightning rod,

> … I say, if these things are so, may not the knowledge of this power of points be use to mankind, in preserving houses, churches, ships etc. from the stroke of lightning,

the focus was on the protection of built structures. At that time, Andre'-Marie Ampere (b. 1775), Georg Simon Ohm (b. 1789), Michael Faraday (b. 1791) and James Clerk Maxwell (b. 1831) were not yet in the world and little was known about the effects of electricity. Coulomb did not publish fundamental work in the field of electricity and magnetism until 1785 to 1789. Processes in the propagation of electricity in the earth, magnetic effects etc. were not yet researched at the time of the invention of the lightning conductor by B. Franklin had not yet been researched.

An intensive discussion of the step and touch voltage emanating from a lightning rod started in Germany since a tragic accident with 4 deaths during a lightning strike into a shelter on a golf course in 2012.

The step and touch voltage during lightning strikes is increasingly becoming the focus of designers of lightning protection systems. The danger is not limited to direct contact with the down conductor. Even when a person approaches a lightning conductor or a tree struck by lightning, there is a risk of electric arcing between the conductor (tree) and the person, the side flash.

The concept of step and touch voltage comes from the 50 Hz world and fits to the processes of earth faults in switchyards and house installations [1, 2]. In switchyards, for example, contact with metal parts is unavoidable when servicing plant components. Contact of a lightning down conductor by a person is a very unlikely event and is a hazard if and only if this conductor carries a dangerous partial lightning current during a lightning strike.

Step voltage has already been extensively reported [11, 18, 34]. In this report, therefore, the main focus in this report is on the touch voltage and the side flash.

In the 50 Hz world, the touch voltage is derived from the ground surface potential alone. When a lightning-current-carrying down conductor is touched, a voltage

J. Meppelink, *The Lightning Rod as a Danger*,
https://doi.org/10.1007/978-3-031-30434-7_1

proportional to the time derivative of the lightning current is induced as a result of the magnetic field in the loop formed at the time of contact. Thus, the galvanic voltage dropping along the down conductor and determined by the transient skin effect, the induced voltage and the voltage dropping above the ground surface between the down conductor entry into the ground and the location of the person act as the touch voltage.

The concept of touch voltage in the classical sense can therefore **not be used** to assess the danger to persons in the case of lightning currents. Instead, in addition to the touch voltage U that occurs, the actual loads that occur due to the electrical energy W and the electrical charge Q acting on the person must also be calculated. Here, $R_P = 1000\ \Omega$[1] is the resistance of the person who is loaded with the current i(t) to earth when touching the down conductor.

$$W = \int_0^\infty R_P i^2(t)dt \quad und \quad Q = \int_0^\infty |i(t)|dt$$

Therefore, the calculation must be carried out in the time domain. A calculation in the frequency domain, as possible for the calculation of the step voltage, is not expedient in the case of the calculation of the touch voltage.

The hazard must be calculated for all three lightning current pulses specified in the standard [4]. The cumulative effect of negative first stroke and negative subsequent stroke must also be evaluated.

In particular, questions are raised about the effectiveness of site insulations using dry and wet insulating materials such as asphalt and water permeable materials for road pavements. From the studies already done [18] on the hazard due to step voltage, the hazard due to the effect of a continuous water layer is already known. This effect must therefore also be taken into account in the case of contact.

An analysis of the international accident statistics for lightning accidents with injury or death shows that the number of injuries or deaths caused by contact with a down conductor is vanishingly small and is not listed in detail at all. About 3–5% of all cases refer to those where a metallic structure is struck by lightning and the resulting potential is transmitted e.g. through water pipes or line-connected equipment (telephone). The risk of injury or death due to touch and step voltages inside the structure and in an area up to 3 m around down conductors outside the structure, as calculated according to EN 62305–2, is vanishingly small. There are no reports from installers of lightning protection systems of injury or death from touch voltages. The vanishingly small number may be due to a person seeking the interior of a building during an approaching thunderstorm. It is extremely unlikely that a person would stand in front of a building with an outstretched arm and touch a bare down conductor during an approaching thunderstorm.

This report shows the means of calculating the touch voltage when a vertical down conductor is touched and the hazards derived from it. Similarly, the hazard

[1] Typical value given in most citations. For more details view IEC 60479-1.

of a person approaching a lightning current carrying down conductor with electrical flashover from the down conductor to the person is investigated. A solution to the problem with insulated down conductors is presented.

Chapter 2
Model for Calculating the Hazard

Figure 2.1 shows a model of the influence and for the calculation of the energy W and the charge Q in a person when touching a down conductor or if the person is standing next to a derivation. The indicated voltage components are effective. The model is assumed for a typical height of contact of 1.4 m and a distance of 0.8 m. The following must be taken into account.

- diameter and material of the down conductor
- height at which a person touches the arrester and distance of the person from the arrester
- insulation of the down conductor and effect of the conductive layer
- site insulation, including multilayer, dry and wet or access restriction
- Effect of water on insulating layers
- Effect of water permeable layers
- Specific resistance and permittivity
- Specific earth resistance, its dependence on geological parameters and environmental influences
- Type of grounding system and potential control
- kC coefficient for the current level in the conductor
- Limits of W and Q for ventricular fibrillation for the three occurring lightning currents
- The electrical strength of the air gap between the person and the down conductor for the occurring voltage waveforms.

Depending on the particular case, the model will be calculated with a network analysis program [9], a program for grounding system analysis XGS_lab [13] and with a numerical field calculation program [14]. From the results, an evaluation is made for different technical solutions and presented in an evaluation matrix in Tables 11.1 and 11.2.

All calculations are performed for the quasi-stationary case and include galvanic, inductive and capacitive coupling in the quasi-stationary range. Electromagnetic waves are excluded from the calculations. For a quasi-stationary calculation, the rise

J. Meppelink, *The Lightning Rod as a Danger*,
https://doi.org/10.1007/978-3-031-30434-7_2

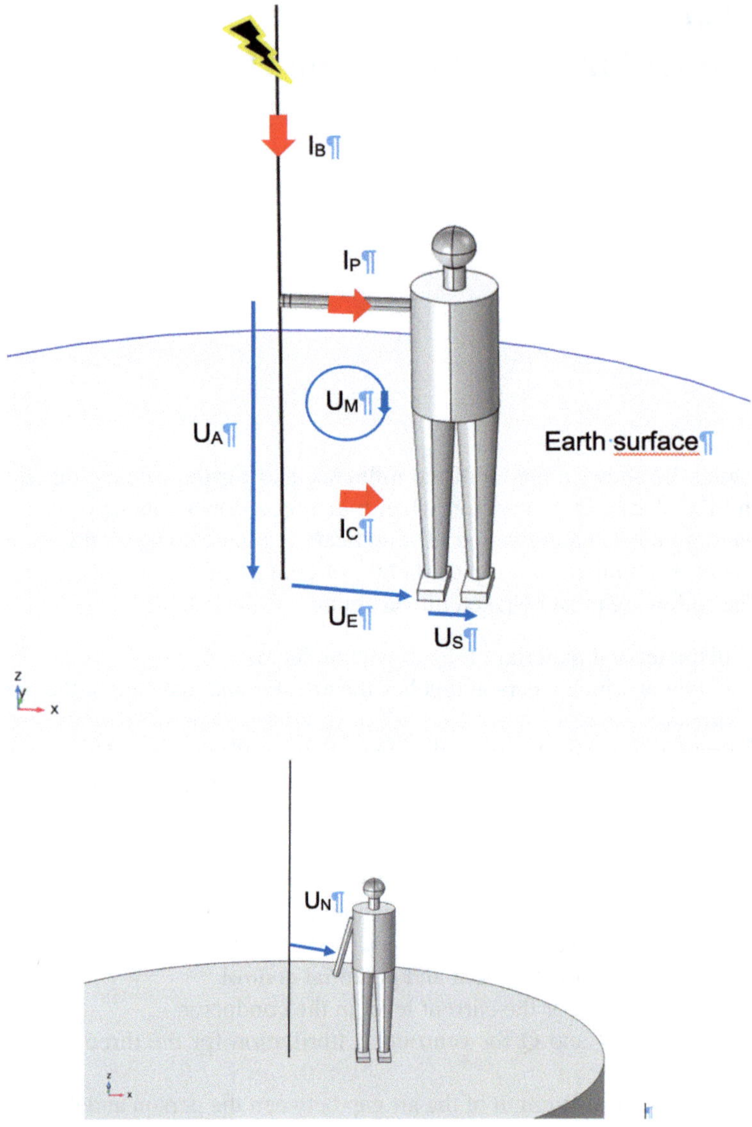

U_A Voltage drop along the down conductor
U_E Voltage drop along the surface of the earth
U_M Voltage induced by the magnetic field
U_N Proximity voltage which determines a side flash
U_S Step voltage
I_B Excitation lightning current
I_P Current in the arm of the person
I_C Capacitive displacement current due to capacitive coupling between live areas and the surface of the person.

Fig. 2.1 Model describing the acting stress components. A person touches the down conductor at a height of 1.4 m and at a distance of 0.8 m as an example

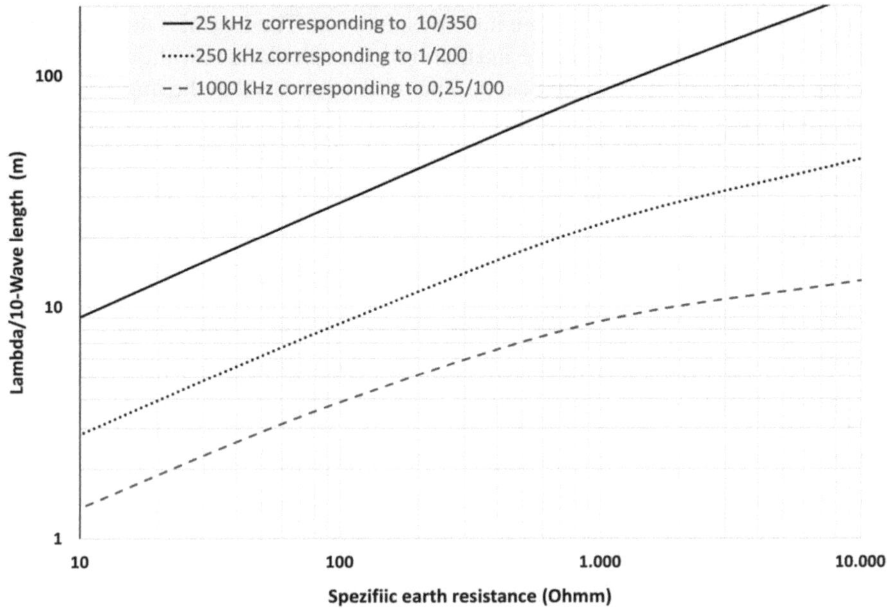

Fig. 2.2 λ/10 Wavelength for the standardized lightning currents (up to this wavelength quasi-stationary calculation is possible) depending on the specific earth resistance for a horizontal or vertical earth electrode. *Source* Meppelink, J.: Measures to reduce the step voltage in grounding systems of lightning protection systems. Final report of Project 83, sponsored by VDE/ABB. April 2020

time of the time domain pulse should be about 10 times larger than the transit time in the system. For the distances in question here to a down conductor of 3 m, a transit time above the earth in air of 10 ns results according to Fig. 2.2. Thus, the condition for a negative sequence current according to Fig. 19.1 with 250 ns rise time is fulfilled. For the time derivative of the follow surge current according to Fig. 19.2, the condition is just fulfilled. Inside soil, the propagation velocity depends on the specific earth resistance and can reach a value of 10% of the speed of light. For the value of the specific earth resistance of 250 Ohmm considered here, a λ/10 wavelength of 5.5 m results for a vertical earth electrode. With an earth electrode length of 9 m in the earth, the condition is then not exactly fulfilled with the lightning current 0.25/100.

With these conditions all of the following calculations are carried out. Unless otherwise emphasized, the analytical lightning currents according to IEC 62305-1 are used as lightning currents. It must be taken into account that the time derivative of the analytical lightning current forms has a 1.38 times larger current slope compared to the normative current slopes specified in IEC 62305-1. The results are primarily presented for lightning protection class I and are also shown for the other lightning protection classes II and III/IV on a case by case basis.

The ionisation in soil is not taken into consideration. The reason is that the ionisation reduces the earthing resistance and therefore the calculations without soil ionisation are to be considered as worst cases. The other reason is that the implementation of soil ionisation increases the CPU time remarkably and the input parameters are not known for a particular case. For further considerations a study in [23] can be helpful.

Chapter 3
Exemplary Buildings with Earthing System Type A and Type B

The exemplary calculation in this report is carried out for a building with dimensions of 15 m × 15 m × 10 m with a type A earth electrode [9], Fig. 3.1. Since the threat is to be evaluated for all 3 lightning current forms, the current distribution to the individual down conductors of an external lightning protection system must be examined. In the standard [12], kc coefficients are given which apply only to the negative subsequence stroke. Therefore, the kc coefficients for the lightning current forms 1/200 and 10/350 were recalculated for the two buildings in Fig. 3.1 and presented in Table 3.1.

The most dangerous case is the contact of a down conductor at the corner of the building. For this purpose, the kc value for the current distribution [12] is determined for all three lightning current forms, Figs. 3.2 and 3.3. It was noticed that a strong dependence of the kc value on the specific ground resistance occurs and the two grounding types A and B show different values. In the present case, however, a earth resistance of $R_A = 10$ Ohm [12] can only be achieved for a specific earth resistance $\rho_E < 250$ Ohmm. Therefore, the k_C values are not applicable for $\rho_E > 250$ Ohmm, in these cases the grounding system must be extended so that a value of $R_A = 10$ Ohm is again achieved. For the determination of the parameters W and Q for the buildings according to Fig. 3.1, the kc values from Figs. 3.2 and 3.3 are used, assuming a specific earth resistance of 250 Ohmm and as a worst case, see Table 3.1.

J. Meppelink, *The Lightning Rod as a Danger*,
https://doi.org/10.1007/978-3-031-30434-7_3

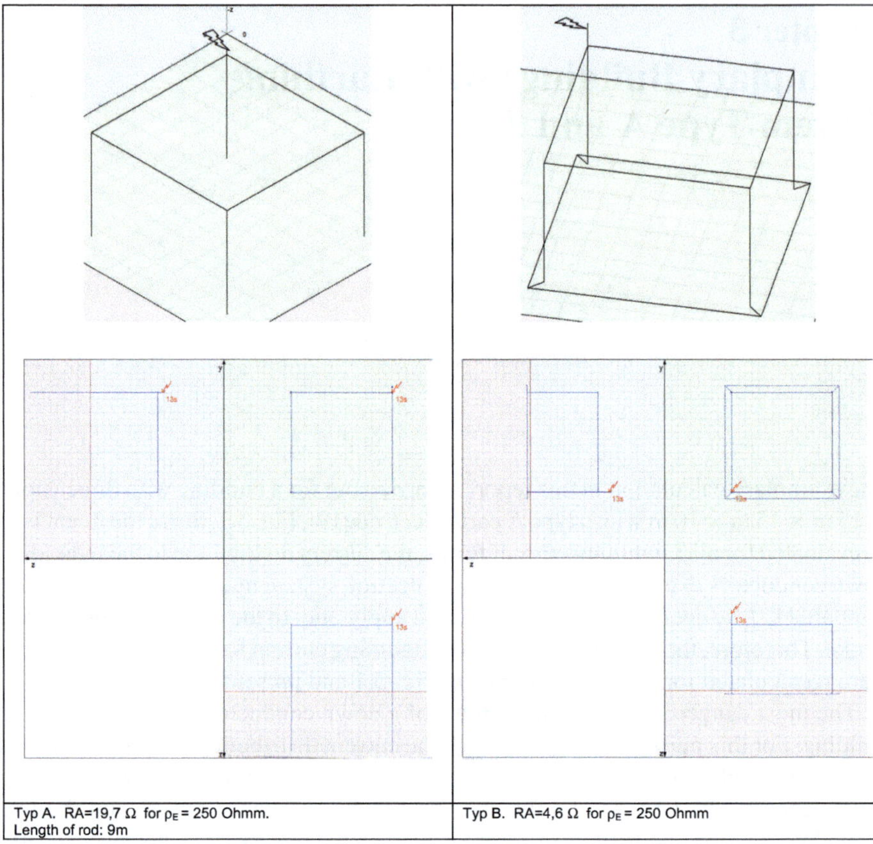

Fig. 3.1 Calculated buildings and earth electrode types type A and type B in 3D and isometric view

Table 3.1 Used k_c values and currents I * k_c for calculation of W and Q for buildings according to Fig. 3.1, Type A and type B and E = 250 Ohmm, Class I

Impulse current [4]	I Class I	kc		I * kc	
		Typ A	Typ B	Typ A	Typ B
	kA			kA	
0.25/100	50	0.494	0.493	24.7	24.65
1/200	100	0.42	0.432	42	43.21
10/350	200	0.26	0.415	52	82.94
	I Class II	kc		I * kc	
		Typ A	Typ B	Typ A	Typ B
	kA			kA	
0.25/100	37.5	0.494	0.493	18.53	18.49
1/200	75	0.42	0.432	31.5	32.4
10/350	150	0.26	0.415	39	62.2
	I Class III/IV	kc		I * kc	
		Typ A	Typ B	Typ A	Typ B
	kA			kA	
0.25/100	25	0.494	0.493	12.35	12.33
1/200	50	0.42	0.432	21	21.6
10/350	100	0.26	0.415	26	41.5

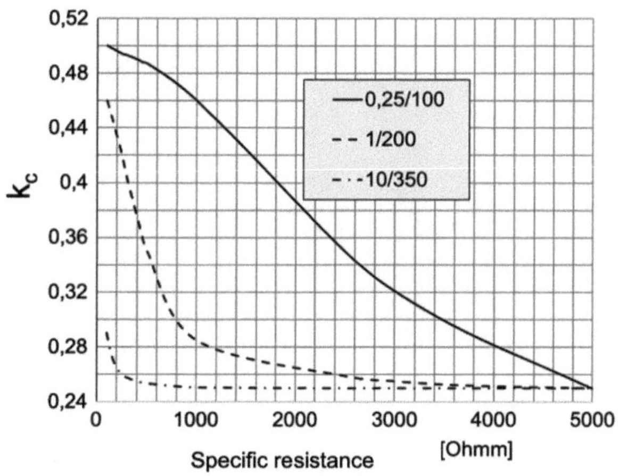

Fig. 3.2 kc-values for earthing system typ A with earth rods of 9 m length

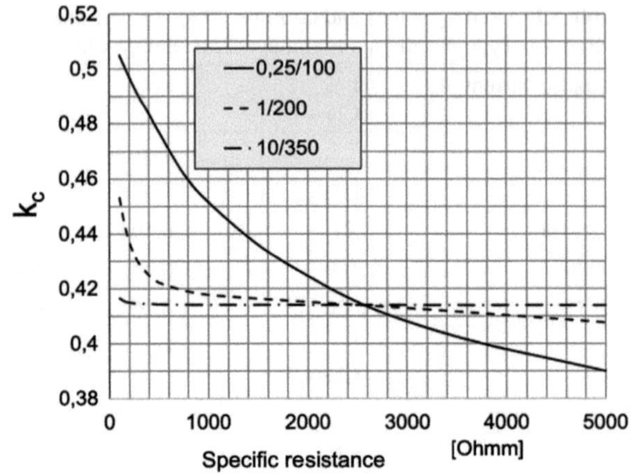

Fig. 3.3 kc-values for earthing system typ B with ring conductor having a size of 4 × 15 m

Chapter 4
Permissible Limits for the Cause of Death Due to Ventricular Fibrillation

4.1 Permissible Limits for Touch Voltage

In the literature [3], a voltage of 2 kV is specified as the permissible touch voltage for the lightning current 10/350. This is only valid for galvanic coupling. In the case of inductive coupling, the specification of the voltage does not make sense, since the voltage form deviates strongly from the 10/350 form. In these cases, the values W and Q must be calculated. Limiting values valid according to the state of the art are named in [3] and are $W_G = 1$ Ws for the energy converted in the body with an equivalent resistance of 1 kOhm and $Q_G = 1$ mAs for the charge. According to the present state of science in this paper, these values should also apply to the lightning currents 1/200 and 0.25/100 [4] and are thus on the safe side.

It is the subject of research whether the limit values valid for the pulse shape 10/350 are also valid for the two other standardized lightning current shapes 1/200 and 0.25/100.

For a load with W and Q values below the limit for ventricular fibrillation, the electrical effect on a person is not described in the literature.

4.2 Limit for Step Voltage

For the step voltage, the current state of the art is a value of 25 kV for pulse shape 10/350 with a body resistance of 1000 ohms between the two feet [34].

J. Meppelink, *The Lightning Rod as a Danger*,
https://doi.org/10.1007/978-3-031-30434-7_4

Chapter 5
Materials

Material data are listed in [2] and shown with the specific resistance. An extract of [2] is shown in Table 5.1. These data can be used for a numerical calculation. The contact resistance of 100 kOhm required in [12] is not specified in the literature. There is also no indication in [12] of the standard according to which this is to be determined.

Table 5.1 Specific resistance of materials for insulating layers [2]

Material	Specific resistance Ohmm		Relative permittivity
	Dry	Wet	
Asphalt	$2 * 10^6 \ldots 30 * 10^6$	$10.000 \ldots 0.2 * 10^6$[a]	2.8–3.3
Concrete	$10^6 \ldots 10^9$	21…200	
Crusher run granite with fines	$140 * 10^6$	1300 Ground water, 45 Ohmm	
0.04 m Crusher run granite with fines	4000	1200 Rain water, 100 Ohmm	
0.025 m bis 0.1 m Washed granite	$1.5 * 10^6$ bis $3 * 10^6$	5000 Rain water, 100 Ohmm	

[a] It can be assumed that the conductivity is reduced by the water film on the surface

J. Meppelink, *The Lightning Rod as a Danger*,
https://doi.org/10.1007/978-3-031-30434-7_5

Chapter 6
Danger from Lightning Currents 10/350, 1/200; 0.25/100 According to Type of Coupling

6.1 Galvanic Coupling Caused by a Down Conductor U_A

For the consideration of the touch voltage, the galvanic coupling voltage U(t) along the surface of a current-carrying cylindrical solid conductor is also to be investigated. A solution for a step function of the current according to [6] is shown in Eq. 6.1. For the case of a lightning current [4], the calculation of the voltage U(t) can be done analytically by convolution or by a network model with a network analysis program. The evaluation of Eq. 6.1 is shown in Fig. 6.1 for the parameters according to Table 6.1

$$\frac{U(t)}{U_0} = 1 + \sum_{i=1}^{\infty} e^{-x_i^2 * \frac{t}{\kappa \mu a^2}} \tag{6.1}$$

$U(t)$ Voltage along the surface of the conductor with a step function of the current
U_0 Voltage with direct current
x_i Zeros of the Bessel function J1(x), [7]
κ Conductivity
μ Permeability
a Radius of conductor
t Time
T Skin time constant $T = \kappa \mu a^2$.

The voltage U(t) in Fig. 6.1 shows a decaying course, with the value for $t = 0$ growing over all limits. In a copper conductor, the current field diffuses gradually into the interior of the conductor and reaches the final value only after about 500 μs. A stainless steel conductor shows a more rapid penetration of the flow field into the conductor due to the smaller skin time constant and reaches the steady-state end value after only 10 μs, but has a DC resistance about 40 times greater than copper.

J. Meppelink, *The Lightning Rod as a Danger*,
https://doi.org/10.1007/978-3-031-30434-7_6

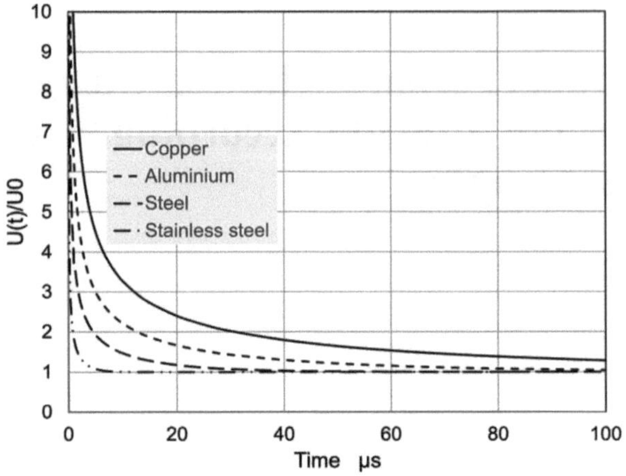

Fig. 6.1 Voltage drop $U(t)/U_0$ referred to the DC current value U_0 along the surface of a cylindrical down conductor with a length of 1 m when excited by a step function of the current according to Eq. 6.1

Table 6.1 Conductivity, temperature coefficients, skin time constant T and DC resistance $R'_{0.20\,°C}$ per meter length for conductor with 1 m length and a = 8 mm diameter

		Aluminium	Stainless steel	Copper	Steel $\mu = \mu_0$
κ	S/m	2.32e7	1.43e6	5.62e7	8.33e6
κ_{cu}/κ		2.42	39.3	1	6.75
α	1/K	3.77e−3	5.3e−3	3.9e−3	4.6e−3
β	1/K^2	1.3e−6	0.6e−6	0.6e−6	6e−6
T	µs	466.2	28.7	1129.4	167.4
$R'_{0.20°C}$	mΩ	0.86	14	0.36	2.4

Because of the sluggish diffusion of the flow field into the interior of the conductor, there is also greater heating of the outer layers. As a result, feedback occurs so that the current density in the outer layer becomes smaller again [38]. The transient skin effect according to Eq. 6.1 can be reproduced by an electrical equivalent circuit of L and R elements according to Fig. 6.2 [8]. Figure 6.3 shows a good agreement of a calculation with Microcap [9] for the same ratios underlying the analytical calculation according to Fig. 6.1.

For the case of a lightning current [4, 8], the loads W and Q were determined with a network analysis program according to Fig. 6.4 and are shown in Table 6.2. For the materials copper, aluminium and steel, the transient skin effect can be neglected. For a down conductor made of stainless steel, the real threat values for the lightning current 10/350 with W = 0.79 Ws and Q = 0.87 mAs already come close to the limit values W_G = 1 Ws and Q_G = 1 mAs, but remain far below the values of the other couplings, which are compiled in Table 6.7 in Sect. 6.8 for the earth electrode type A. The transient skin effect can therefore be neglected for the further calculations performed here.

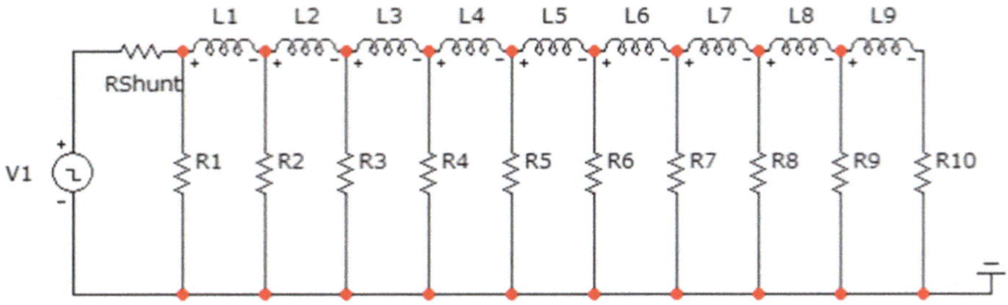

L1-L9 Partial inductance of the down conductor
R1-R9 Partial resistance of the down conductor of stainless steel

Fig. 6.2 Simulation model for the transient skin effect in the cylindrical solid conductor of 8 mm diameter

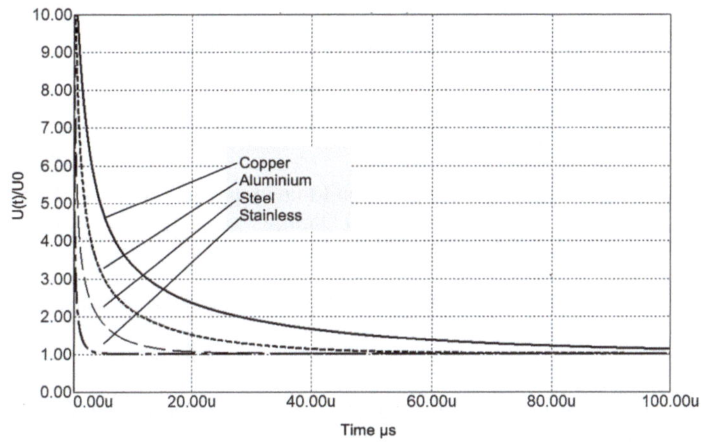

Fig. 6.3 Voltage drop referred to the DC case from the simulation according to Fig. 6.2 when energized by a step function of the current. Length: 1 m

6.2 Galvanic Coupling U_E Due to Lightning Current in Soil

The galvanic coupling U_E occurs due to the voltage drop between the entry of the discharge into the earth and the location of the person (foot) along the earth surface in Fig. 2.1. U_E was determined by a numerical field calculation [14] in the time domain in a model according to Figs. 6.5 and 6.6 for an earth resistivity of $\rho_E = 250$ Ohmm for the negative sequence lightning with data according to Table 3.1.

The person in Fig. 6.6 was represented according to Fig. 6.5 such that the total resistance of the body between the point of contact at the interception rod and the bottom of the two feet is 1000 ohms. This value is decisive for the calculation of the touch voltage, which is in the foreground here.

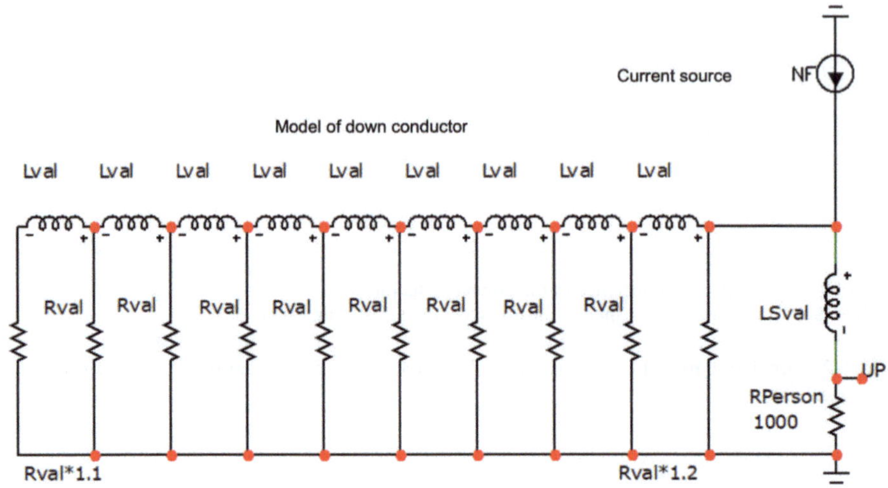

Lval	Partial inductance of the down conductor 12.5 nH	12,5 nH
Rval	Partial resistance of the down conductor stainless steel 140 m	140 mΩ
LSval	Circular inductance of the build-up loop 3.32 μH	3,32 μH
RPerson	Simulation of a person 1000 Ohm	1000 Ohm
UP	Potential at the point of contact	

Fig. 6.4 Network for calculating the energy W and charge Q converted by transient skin effect in a person

Table 6.2 W and Q due to the transient skin effect in the down conductor, calculated for values in Table 3.1. Person touched at a height of L = 1.4 m

	Copper 8 mm		Aluminium 8 mm		Steel 8 mm		Stainless steel 8 mm	
	W	Q	W	Q	W	Q	W	Q
	Ws	mAs	Ws	mAs	Ws	mAs	Ws	mAs
10/350	0.001	0.022	0.004	0.053	0.025	0.15	0.79	0.87
1/200	275e−6	0.006	921e−6	0.016	0.004	0.045	0.126	0.264
0.25/100	75e−6	0.002	233e−6	0.0046	0.001	0.012	0.021	0.075

In the course of the investigations, an effect of the step voltage was later determined despite the centre distance of the feet of only 20 cm. The evaluations of the step voltages in individual sections are to be regarded as orientational, especially because the step width in the model is only 0.2 m. The results of the investigations are based on the results of the test. If a more precise determination of the normative step voltage is the focus, the use of standard programs such as XGS-Lab Grounding System Analysis [13] is recommended.

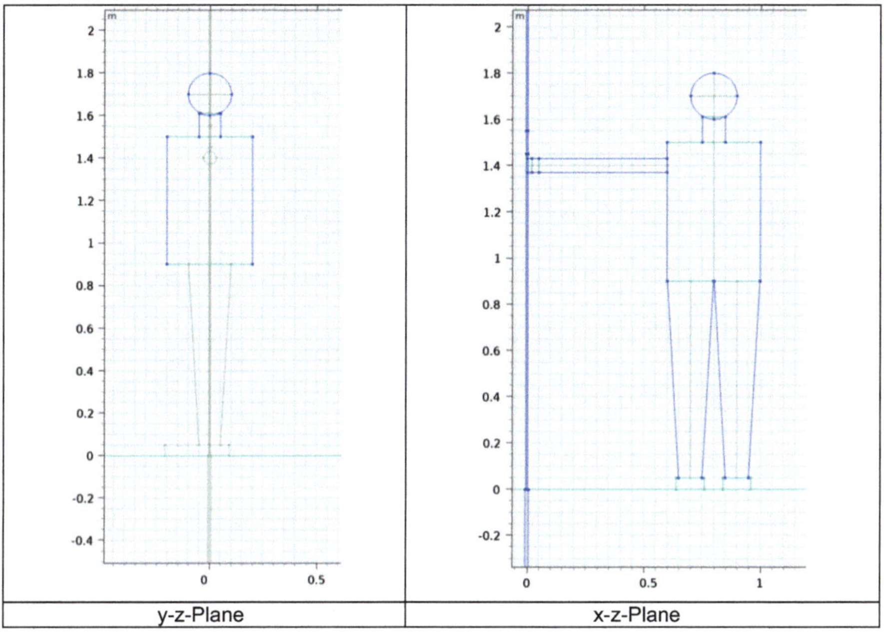

Fig. 6.5 Geometry of the person model according to Fig. 2.1 for numerical field calculation

To limit the CPU time, the two cylinders b and c were introduced and provided with a finer meshing, as shown in Fig. 6.7. The calculation was performed for the negative sequence current 0.25/100 as an example.

The galvanic coupling is first calculated for the case of contact of a person at the down conductor. With regard to the use of an isolated down conductor as a problem solution, the case without contact of the person at the down conductor is then calculated in Sect. 6.4. This calculation then results in the required withstand voltage of an insulated down conductor.

6.2.1 Galvanic Coupling U_E. Person in Contact with a Down Conductor

Figure 6.8 shows the current flow through the person and in the earth. The current field in the earth is influenced by the current flow in the person. Therefore, the voltage U_E cannot be calculated conventionally from the course of the earth surface potential, which is easy to calculate. The potential curve along the line running on the earth's surface in Fig. 6.9 in Fig. 6.10 shows the influence of the person. The current flowing through the person can be determined by integrating the current density in the arm or body with the functions within the calculation program, and can be used to evaluate

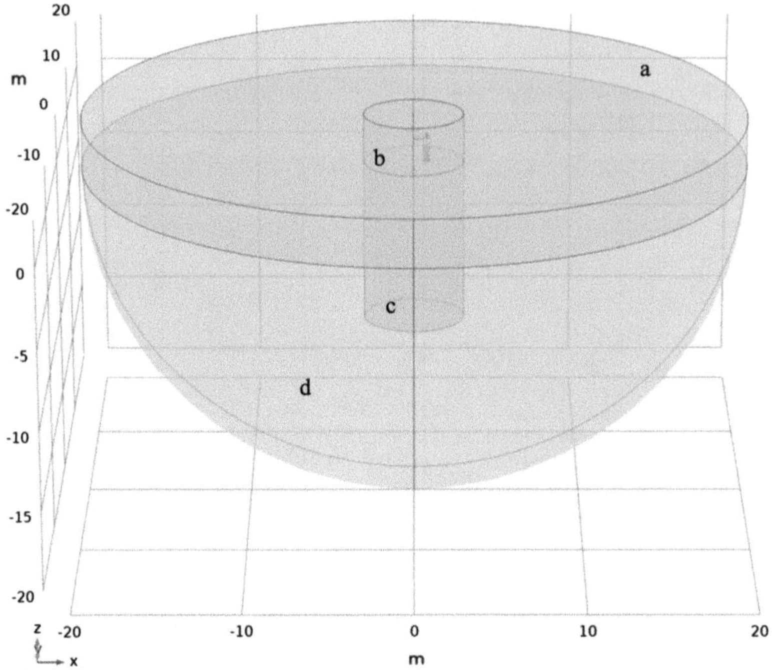

a Airspace above ground
b,c Volume with refined meshing
d Earth

Fig. 6.6 Geometry of the model according to Fig. 2.1 for numerical field calculation with a person shown in Fig. 6.5

the hazard values W and Q. The calculation for the lightning currents 1/200 and 10/350 was also performed and shows basically the same behaviour, so only the results are presented in Table 6.7 in Sect. 6.8. The values of galvanic coupling due to the current in the ground massively exceed the limit values WG and QG. The lightning current of the form 10/350 represents the greatest danger, followed by the 1/200. The negative sequence lightning 0.25/100, on the other hand, is decisive for the inductive coupling, see Sect. 6.3.

6.2.2 *Galvanic Coupling U_E. No Contact of a Person with the Down Conductor*

The calculated arrangement corresponds to Fig. 6.6, where the person with the outstretched arm does not touch the discharge. Figure 6.11 shows the corresponding field image. Only a small capacitive current of 3 A flows through the arm, see

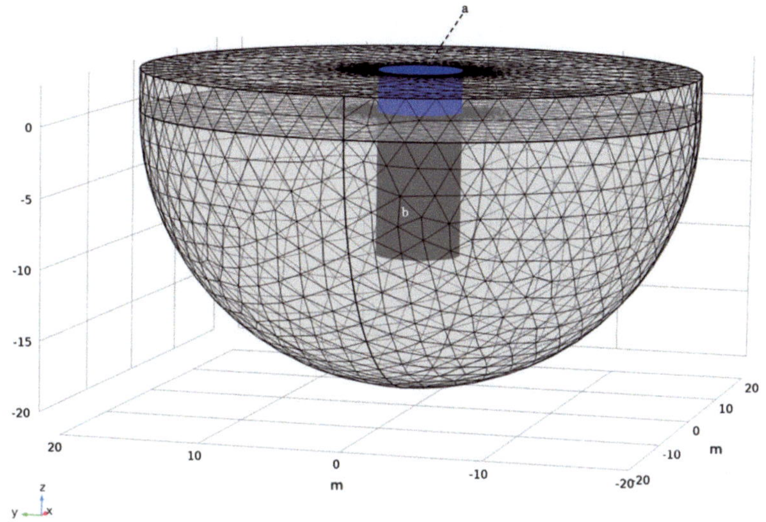

Fig. 6.7 Meshing of the areas

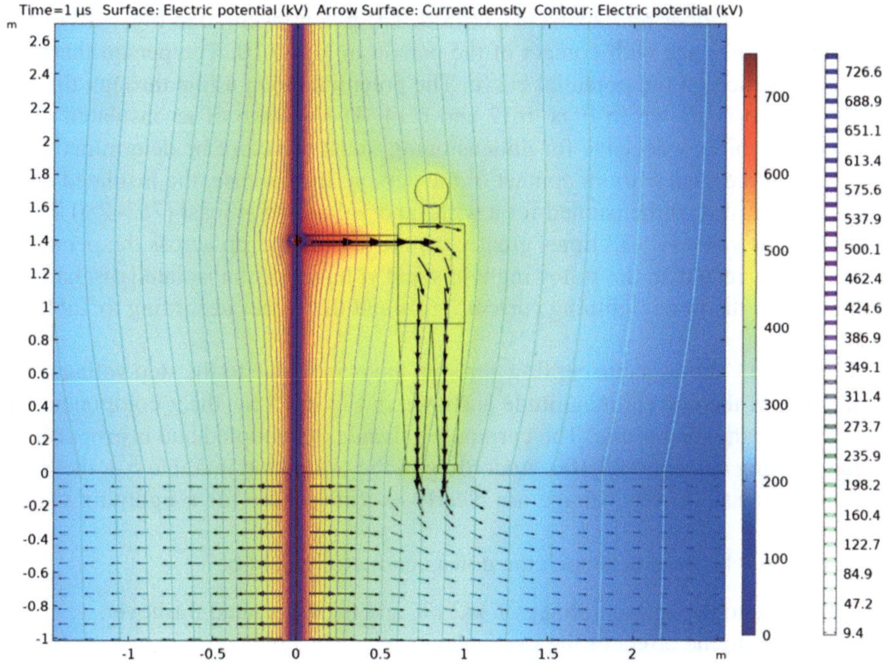

Fig. 6.8 Field image at time of current peak at t = 1 μs with equipotential surface and lines and vectors of current density. Negative following flash 0.25/100 with a peak value of 24.7 kA according to Table 3.1. Soil ρ_E = 250 Ohmm. Current density vectors in logarithmic representation

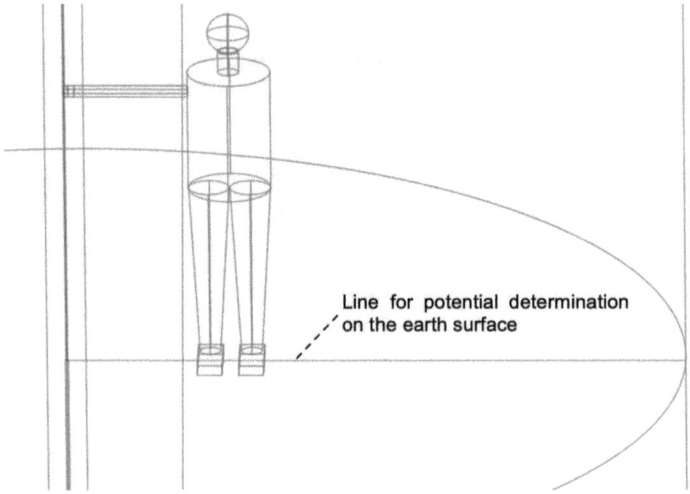

Fig. 6.9 Line for potential determination on the earth surface

Fig. 6.16, which does not represent a hazard. The potential curve in Fig. 6.12 differs from that in the case with contact of the person in Fig. 6.10. The person therefore has an influence on the potential curve. The potential along a line through the arm of the person is shown in Figs. 6.12 and 6.14. The voltage of an insulated down conductor, which is decisive for dimensioning, must therefore be determined from Fig. 6.14 in the case without contact of the person. In this case, the insulated down conductor must be dimensioned for a withstand voltage of at least (782–280 kV) = 482 kV. This value is 1.33 times greater than the voltage drop across the person on contact. This results in the following withstand voltages for an isolated discharge in Table 6.3 for the three lightning current forms and the values according to Table 3.1 (Fig. 6.13).

Figure 6.11 shows a current flow through the person due to the step voltage. The evaluation of the current magnitude is shown in Fig. 6.15 for the z-component and Fig. 6.16 for the magnitude. The current is galvanically coupled and is proportional to the exciting current 0.25/100. According to the evaluation in Table 6.4, the limits for the permissible step voltage are clearly exceeded here for a lightning current 10/350.

A person who does not touch the down conductor would be

(a) Die due to the acting voltage of 482 kV and the associated flashover.
(b) Die due to the effect of the step voltage.

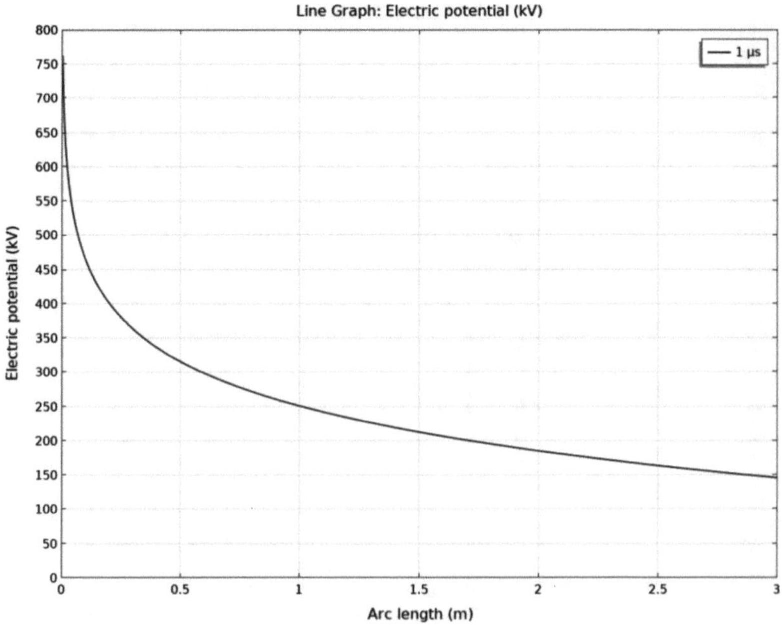

Without person on the earth's surface

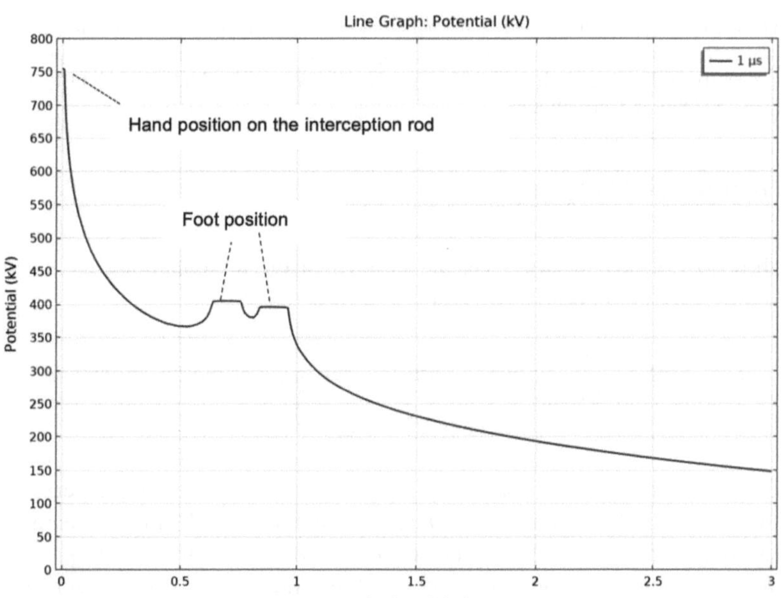

With person on the earth surface

Fig. 6.10 Potential curve on the earth's surface along line B in Fig. 6.9 at the time of the current peak $t = 1\ \mu s$ for 0.25/100

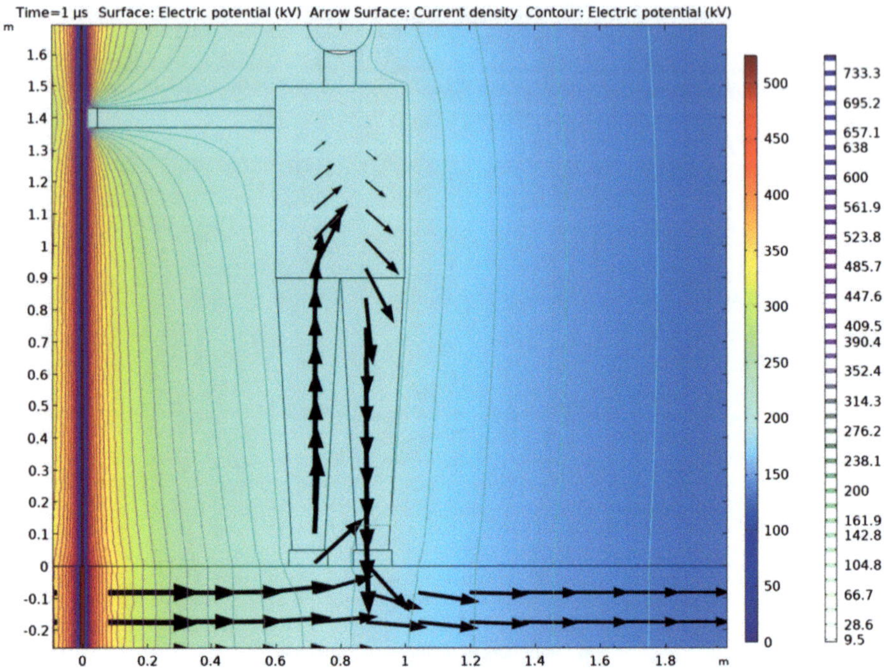

Fig. 6.11 Field image for the case when the person does not touch the down conductor. Field image at the time of the current peak at t = 1 μs with equipotential surface and lines and vectors of current density. Negative following flash 0.25/100 with a peak value of 24.7 kA according to Table 3.1. Soil: $\rho_E = 250$ Ohmm. Current density vectors in logarithmic representation

6.3 Inductive Coupling U_M

According to Fig. 2.1, a stress U_M is induced in the area spanned by the person between the derivative, arm, body and foot point.

$$U_M = M * \frac{di(t)}{dt} \text{ with } M\text{: mutual inductance of the arrangement, see Sect. 17.2.2, Fig. 17.7} \quad (6.2)$$

The effect of the induced voltage by the energy W converted in the person with a resistance of 1000 Ω and the charge Q can be calculated with simple means using a network analysis program [9]. The surface on which the person is standing is assumed to be the equipotential surface. In the network in Fig. 6.17, U_M is represented by the controlled source V(M). Here, the current in the person can be neglected compared to the injected lightning current. Furthermore, the circuit inductance $L_{Circuit}$ has to be considered, which can be determined elementarily according to the literature [15]. Figure 6.17 shows a complete equivalent circuit for the determination of the physical quantities as potential U_P, energy W and charge Q acting on a person and will be

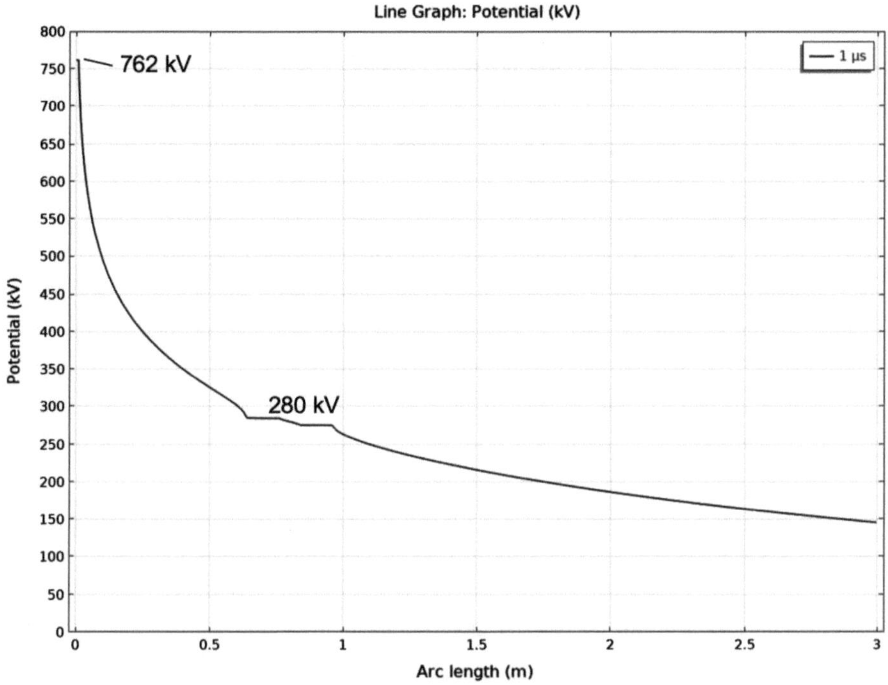

Fig. 6.12 Potential curve on the earth's surface along line B in Fig. 6.9 at the time of the current peak t = 1 μs for 0.25/100

Table 6.3 Results for the case without touching the down conductor. Type A earth electrode; Class I

	U_P	U_S	U_S/U_P
	kV	kV	
10/350	767	1000	1.30
1/200	621	814	1.31
0.25/100	362	482	1.33

Table 6.4 Current level IK at step voltage. Type A earth electrode; Class I

	I_B	I	Limit value according to [34]
	kA	A	A
10/350	52	71.6	25
1/200	42	57.8	No limit value specified. The limit value for 10/350 is adopted
0.25/100	24.7	34	

I_B Peak value of the injected lightning current according to Table 3.1
I Peak value of the current in the body
U_P Voltage between hand and foot
U_S Impulse withstand voltage for one isolated conductor

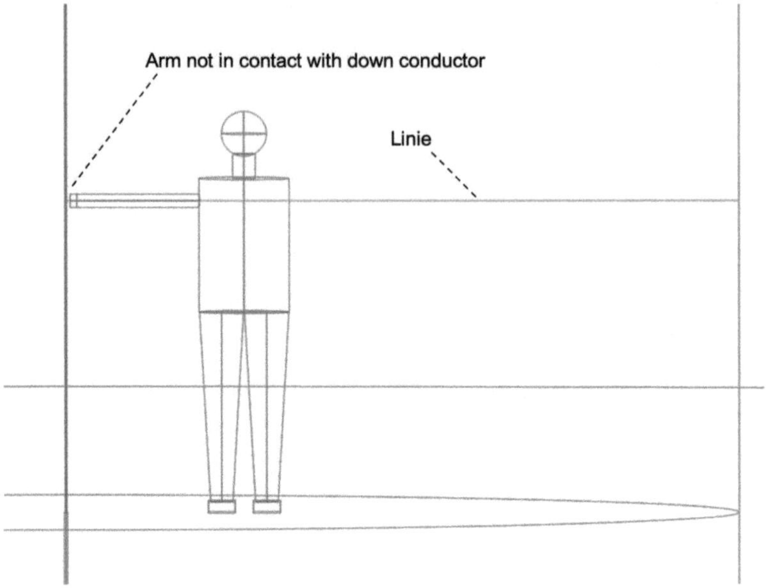

Fig. 6.13 Line for calculating the potential along the arm of the person

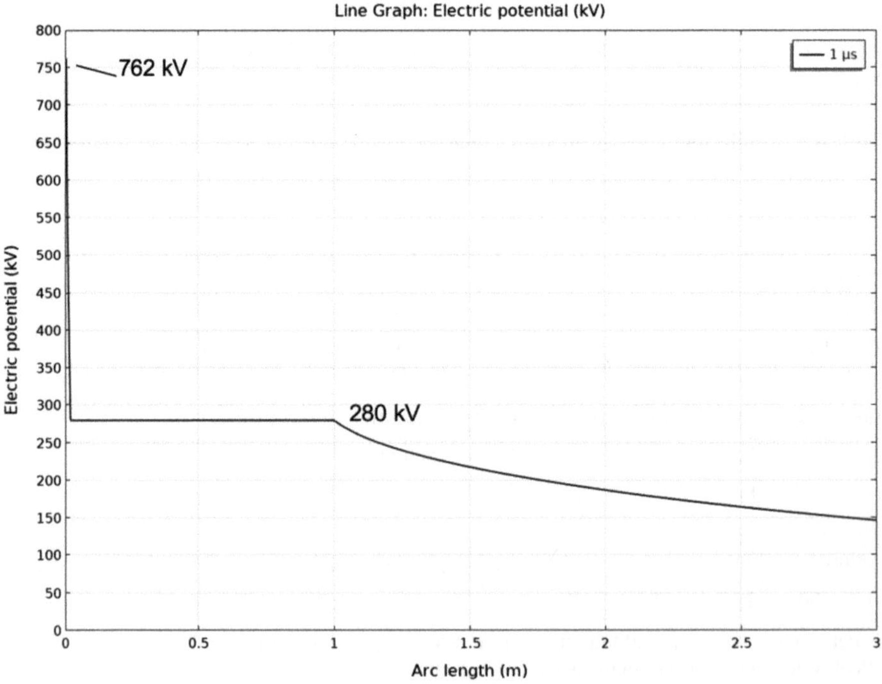

Fig. 6.14 Potential curve along line in Fig. 6.13 at the time of the current peak t = 1 µs for 0.25/100

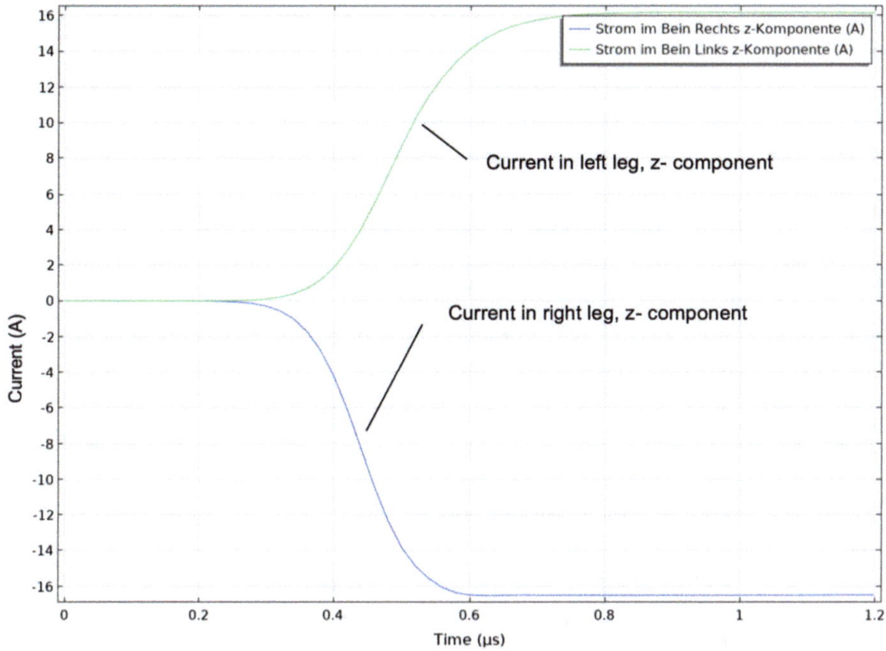

Fig. 6.15 Current curve (ec.Jz) in both legs for 0.25/100

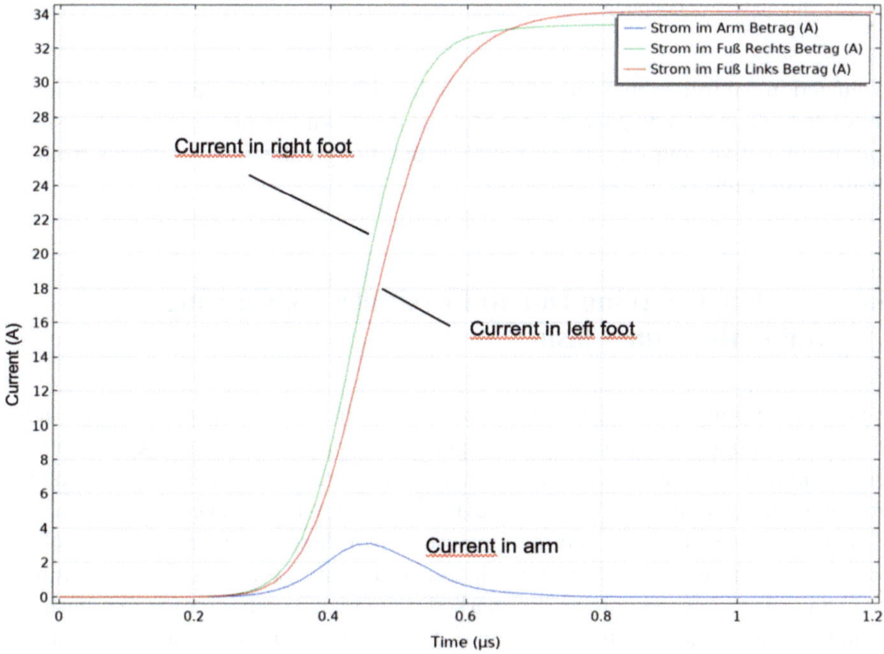

Fig. 6.16 Current curve (magnitude) in both legs for 0.25/100

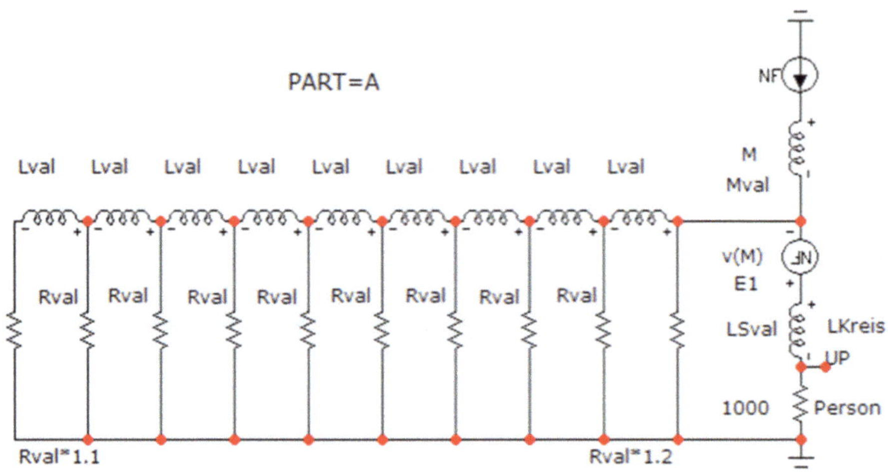

Lval	Partial inductance of the down conductor	12,5 nH
Rval	Partial resistance of the down conductor	140 mΩ
NF	Stainless steel	
M	Lightning current source for impressed current according to IEC 62305-1	1,48 µH
LKreis	Inductance of the build-up loop	3,32 µH
Person	Inductance of the build-up loop	1000 Ω
UP	Potential at the point of contact	

Fig. 6.17 Equivalent circuit diagram for calculating the touch voltage and the values W and Q with galvanic and inductive coupling

evaluated in the following section. The graph in Fig. 6.42 shows the values W and Q converted in a person by galvanic coupling of the down conductor (negligibly small) and the inductive coupling. Here it is clear that the greatest danger comes from the lightning currents 0.25/100 and 1/200.

6.4 Electric Coupling Due to Proximity U_N Causing a Possible Side Flash

A known coupling occurs, in contrast to contact, already when a person approaches the down conductor. It is known from personal injury cases that a person seeking shelter under a tree was killed by a side flash. Therefore, strictly speaking, the concept of touch voltage applies only when the down conductor is touched.

The most probable case and the greatest danger, however, is the presence of one or more persons around the down conductor (proximity), see Fig. 17.1. According to the data in Fig. 2.1, the voltage U_N marked there occurs, which leads to a flashover from the down conductor to the person if the magnitude of the voltage U_N is greater than the electrical strength U_d of the section. The voltage U_N is determined by two effects:

(a) inductive coupling, described in Sect. 6.3.
(b) Galvanic coupling, described in Sect. 6.2.

It is therefore simplified that $U_N = U_M + U_E$. However, according to the data in Sect. 6.8, Fig. 6.40, the maximum voltage occurring is proportional to the course of the exciting current and the following applies: $U_{N,max} = U_{E,max}$ and $U_N = U_E$. This means that the galvanic coupling is determining. The voltage U_N determining a possible flashover can therefore be determined from the course of the potential at the prospective contact point at the discharge (1.4 m height) and the potential on the earth's surface at the position of the foot with the numerical field calculation. Two distances have been calculated for this purpose. At the original distance, the field image with the x-component of the electric field strength is shown in Fig. 6.20. Because of the continuity of the tangential component, the electric field runs parallel to the earth's surface and decays in magnitude with increasing distance from the down conductor as shown in Fig. 6.23. The potential along a line through the body at the height of the arm according to Fig. 6.21 is shown in Fig. 6.22. The same calculation was performed for a distance increased by 3 m, see Figs. 6.24, 6.25, 6.26 and 6.27.

In the following, the arrangement with the person shown in Fig. 6.18, in addition to a derivation, is approximated by a rod-plate arrangement known in the literature. The 50% breakdown voltage [55] of such a rod-plate arrangement $U_{d,50\%}$ can be calculated for a rod-plate arrangement according to Eq. 6.3a as given in [25, 56]. For this purpose, it is necessary to know the mean streamer gradient, which is given in [56] according to Table 6.5. The resulting impulse withstand voltages $U_{d,0\%}$ according to Eq. 6.3b are shown in Table 6.5 for dry conditions and under rain [57, 58] and in Fig. 6.19. For negative polarity, the withstand impulse voltage is larger than for positive polarity. Therefore, positive lightning is the greatest hazard.

$$U_{d,50\%} = E_S * d \tag{6.3a}$$

$$U_{d,0\%} = U_{d,50\%} - 3 * \sigma \tag{6.3b}$$

$U_{d,50\%}$ Breakdown voltage with a probability of 50.
$U_{d,0\%}$ Impulse withstand voltage with a probability of 0%
E_S Mean streamer gradient at $U_{d,50\%}$
σ Standard deviation (Typical value for air spark gaps: 3% of $U_{d,50\%}$
d Gap distance.

The dangerous areas in which $U_N > U_{d,0\%}$ are determined using the data in Table 6.5 and the occurring stresses U_N according to Figs. 6.22 and 6.26. These areas are shown in Figs. 6.28, 6.29 and 6.30 for the three lightning protection classes. From this, the necessary safety distance of a person to a lightning-current-carrying down

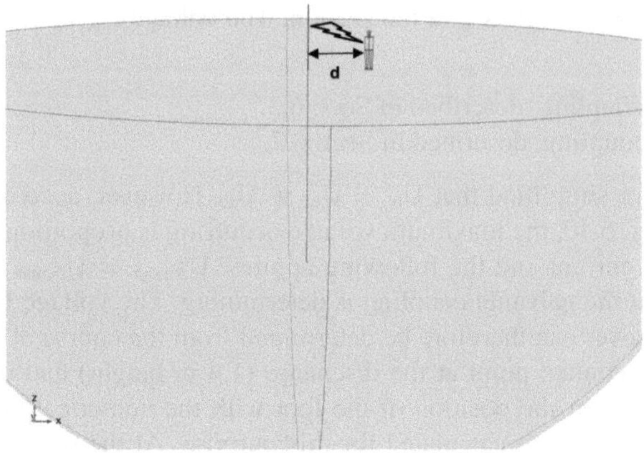

Fig. 6.18 Person in the danger zone for a flashover from the discharge to the person with the distance d

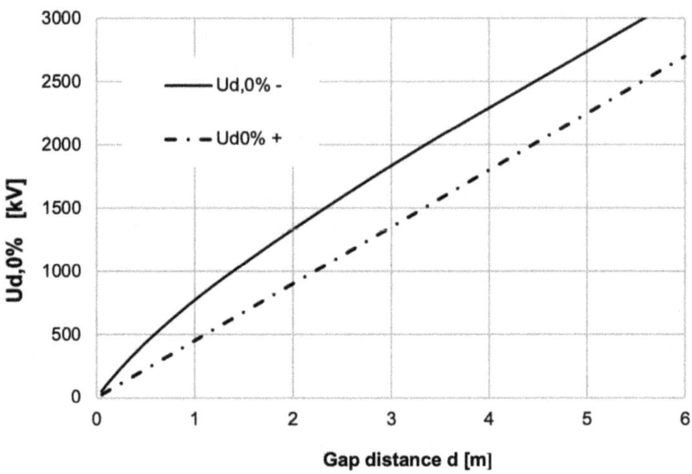

Fig. 6.19 Impulse withstand voltage $U_{d,0\%}$ for a rod-plane arrangement with the gap distance d according to Eq. 6.3a

conductor can be derived. Only the influence of rain is taken into account at 10%. No further safety factors are included.

The voltage U increases proportionally to the specific earth resistance. The dangerous area then increases approximately proportionally.

Since no negative subsequent stroke occurs without a preceding negative initial flash, a plot for the negative subsequent stroke is not required. The greatest hazard in field coupling by proximity is the positive flash of the form 10/350.

During a flashover, the energy W and the charge Q are converted in one person. At a distance of 0.8 m, the values shown in Table 6.7 apply. At distances >0.8 m,

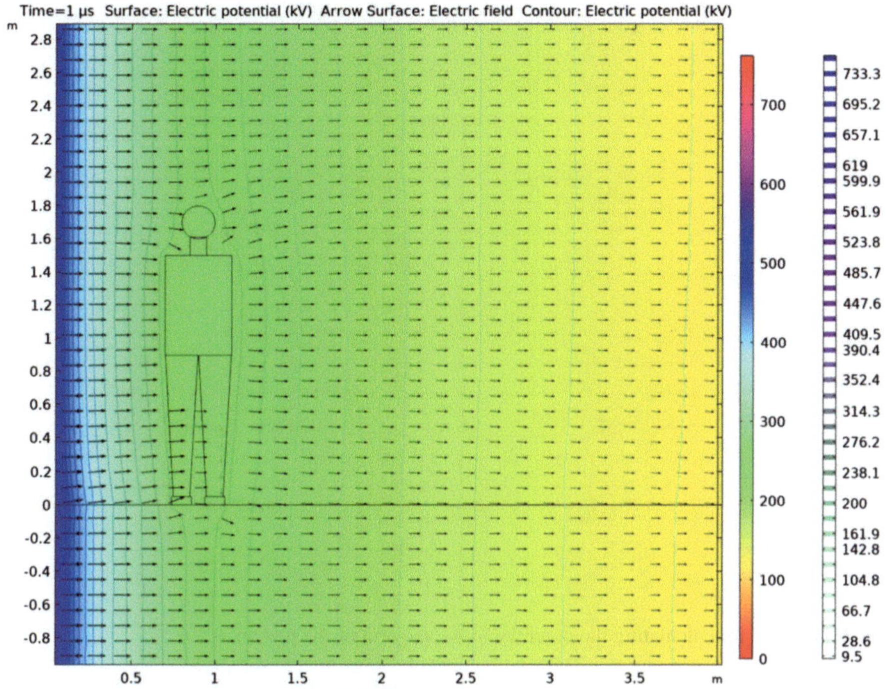

Fig. 6.20 Field image for the case when the person does not touch the down conductor (in the initial position at a distance of 0.8 m). Field image at the time of the current peak at t = 1 µs with equipotential surface and lines as well as vectors of the electric field. Negative subsequent stroke 0.25/100 with a peak value of 24.7 kA according to Table 3.1. Soil $\rho_E = 250$ Ohmm. Field strength vectors in logarithmic representation

these values increase. At distances <0.8 m, the values become smaller according to the potential curve, see Figs. 6.22 and 6.26. However, since the potential curve drops off very steeply and the values in Table 6.7, line a) indicate a massive exceeding of the limit values, a detailed calculation of the W and Q values as a function of the distance of the person to the discharge is omitted.

6.5 Capacitive Coupling

Capacitive coupling is a quasi-stationary coupling and occurs at large voltage slopes, as in the 0.25/100 and 1/200 lightning currents. Capacitive coupling is included in the galvanic coupling studies and is presented and evaluated in these sections.

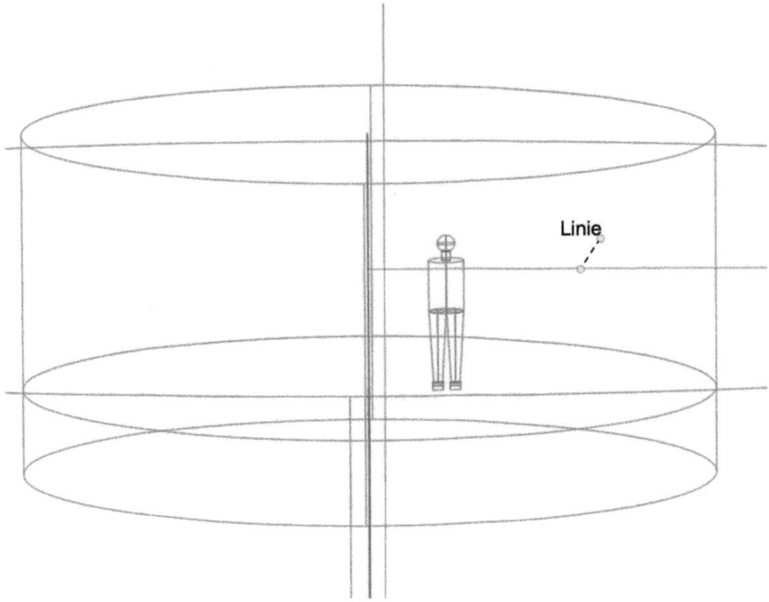

Fig. 6.21 Line for calculating the potential along the arm of the person

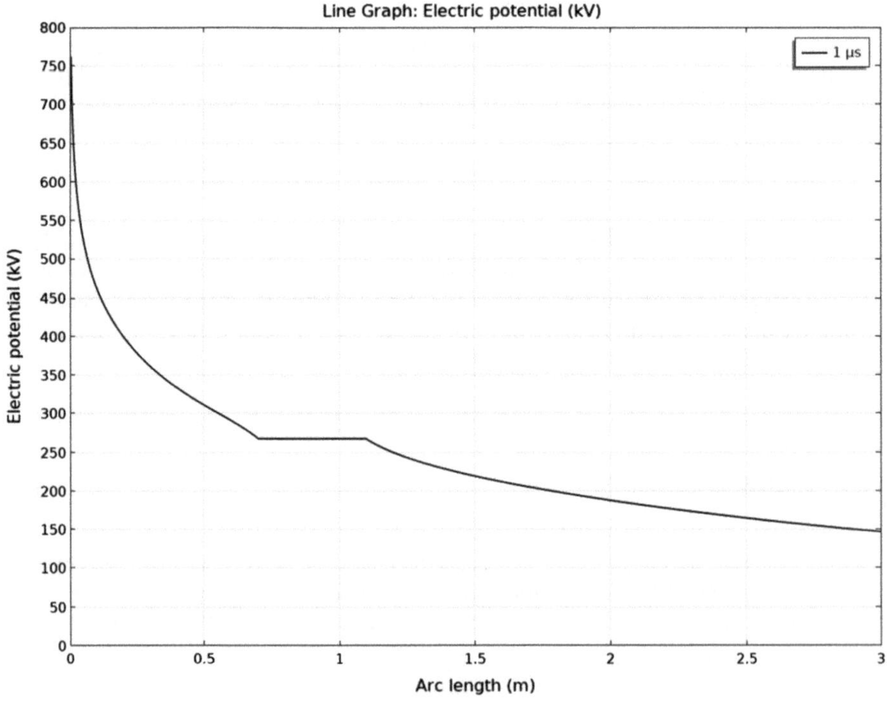

Fig. 6.22 Potential curve along line in Fig. 6.21 at time of current peak t = 1 μs for 0.25/100

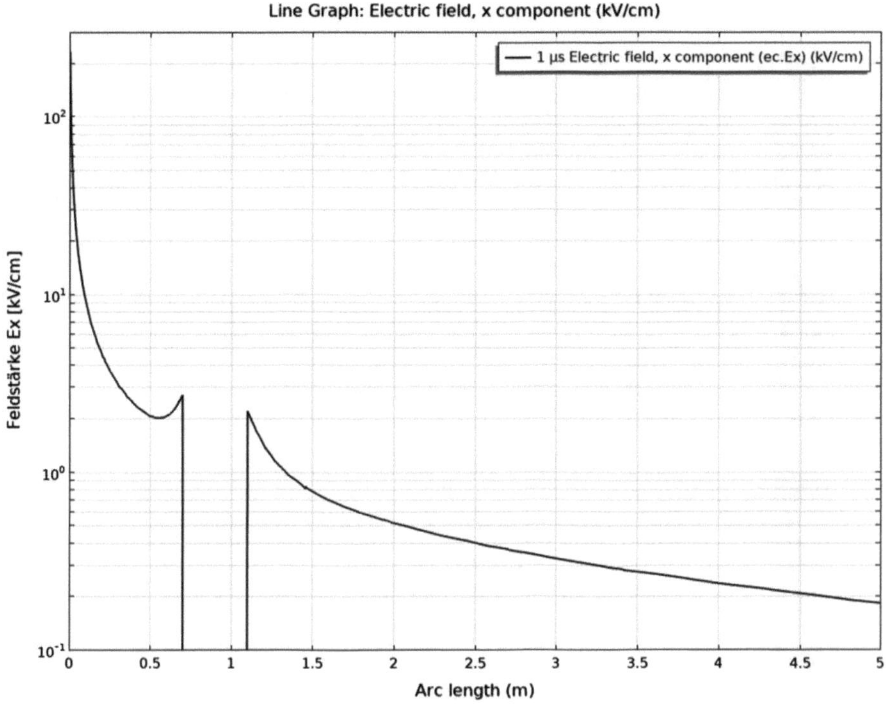

Fig. 6.23 Course of the x-component of the electric field strength along line in Fig. 6.21 at the time of the current peak t = 1 μs for 0.25/100

6.6 Electromagnetic Radiation

Radiation coupling is a transient coupling and requires the complete solution of the wave equation. Limit values by for the energy and charge acting in a person by radiation coupling are currently unknown. The study of radiation coupling is the subject of further research.

6.7 Influence of Equipotential Control Measures

In order to avoid the hazards, it is subsequently examined whether a potential control in the ground is expedient to reduce the galvanic coupling. According to Fig. 6.31, the potential control is designed as a mesh grid with a mesh width of 0.5 m and 6 mm stainless steel rods and is located at a frost-proof depth of 0.5 m. The potential control is then installed in the ground.

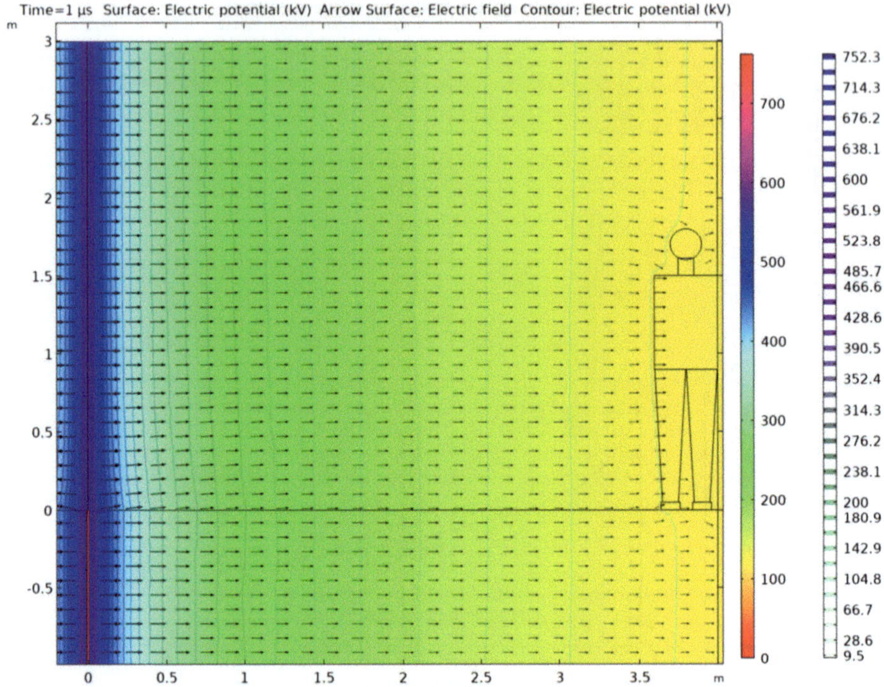

Fig. 6.24 Field image for the case when the person does not touch the down conductor (in the initial position at a distance of 3.3 m). Field image at the time of the current peak at t = 1 µs with equipotential surface and lines as well as vectors of the electric field. Negative following flash 0.25/100 with a peak value of 24.7 kA according to Table 3.1. Soil $\rho_E = 250$ Ohmm. Field strength vectors in logarithmic representation

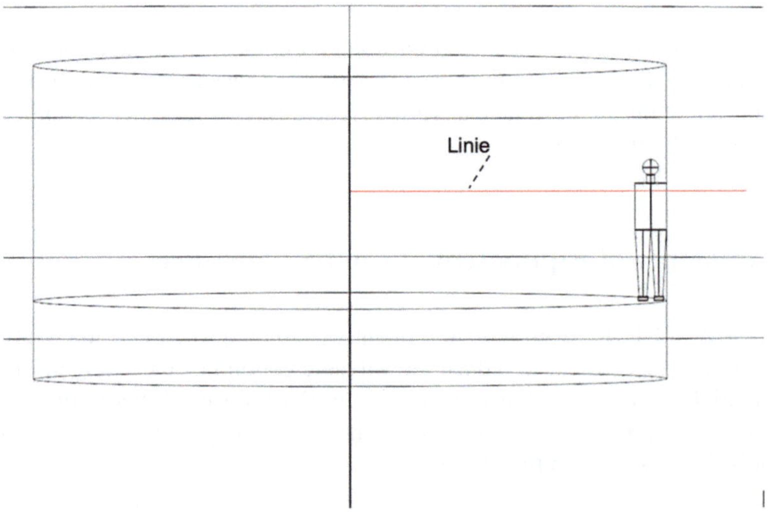

Fig. 6.25 Line for calculating the potential along the arm of the person

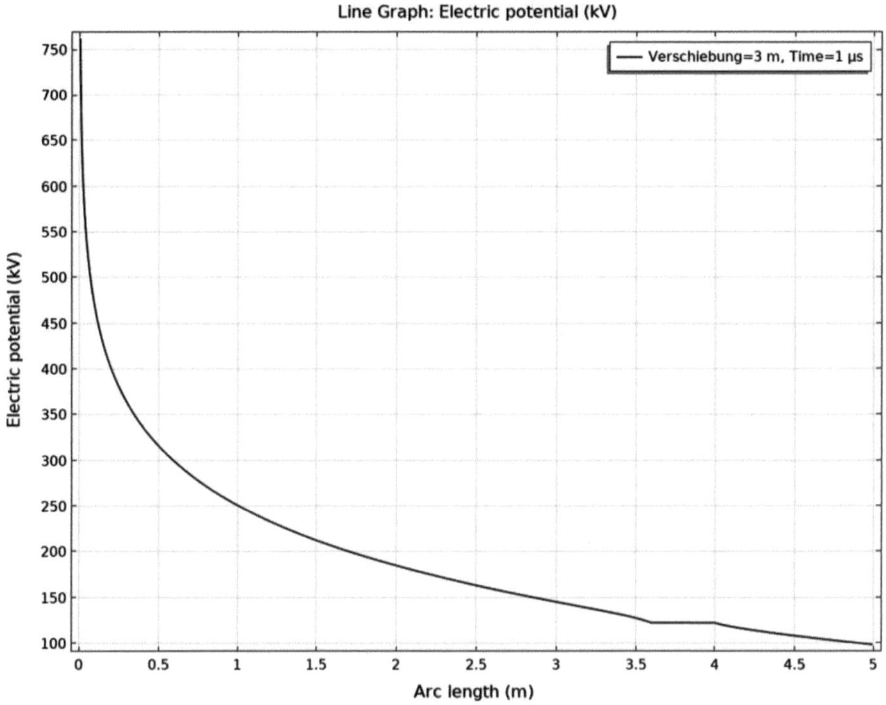

Fig. 6.26 Potential curve along line in Fig. 6.25 at time of current peak t = 1 μs for 0.25/100

Table 6.5 Average streamer gradient E_S at $U_{d,50\%}$ and withstand impulse voltages Ud0%

	Dry						Rain (Faktor 0.9)	
d	E_{S+}	E_{S-}	$U_{d,50\%+}$	$U_{d,50\%-}$	$U_{d,0\%+}$	$U_{d,0\%-}$	$U_{d,0\%+,rain}$	$U_{d,0\%-,rain}$
m	kV/cm		kV					
0.5	5	10.5	250	525	227.5	477.8	204.75	429.98
1		9.5	500	950	455	864.5	409.50	778.05
2		8	1000	1600	910	1456	819.00	1326.78
3		7.5	1500	2250	1365	2047.5	1228.50	1842.75
4		7	2000	2800	1820	2548	1638.00	2293.20
5		6.7	2500	3350	2275	3048.5	2047.50	2743.65
6		6.5	3000	3900	2730	3549	2457.00	3194.10

6.7.1 *Galvanic Coupling U_E with a Person in Contact with Down Conductor in Case of Equipotential Control*

The field image of the current density in Fig. 6.32 shows a significant reduction of the stress on the person compared to Fig. 6.8. As expected, the potential curve along the

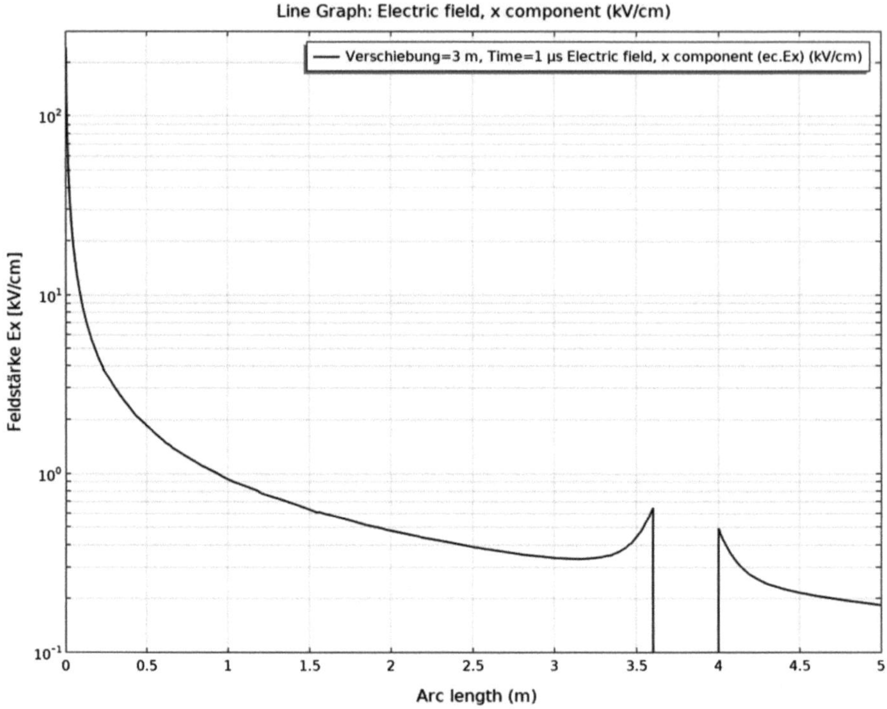

Fig. 6.27 Course of the x-component of the electric field strength along line in Fig. 6.25 at the time of the current peak t = 1 μs for 0.25/100

earth's surface is much flatter compared to that in Fig. 6.10. The current and voltage values are significantly reduced and evaluated in Table 6.6 for the other lightning current forms as well. Nevertheless, the values for energy W and charge Q in the person exceed the limit values. Surface discharges do not occur here because the electric field strength between the foot of the person and the equipotential bonding grid is too small. Because the limit values are exceeded by galvanic coupling, the inductive coupling is not calculated here. The effect of inductive coupling corresponds to the values in Table 6.7b). Measures to reduce the energy and the charge are described in Chap. 12 (Fig. 6.33).

The conclusion is: For the touch voltage there is still the risk to die even when a potential control measure is applied.

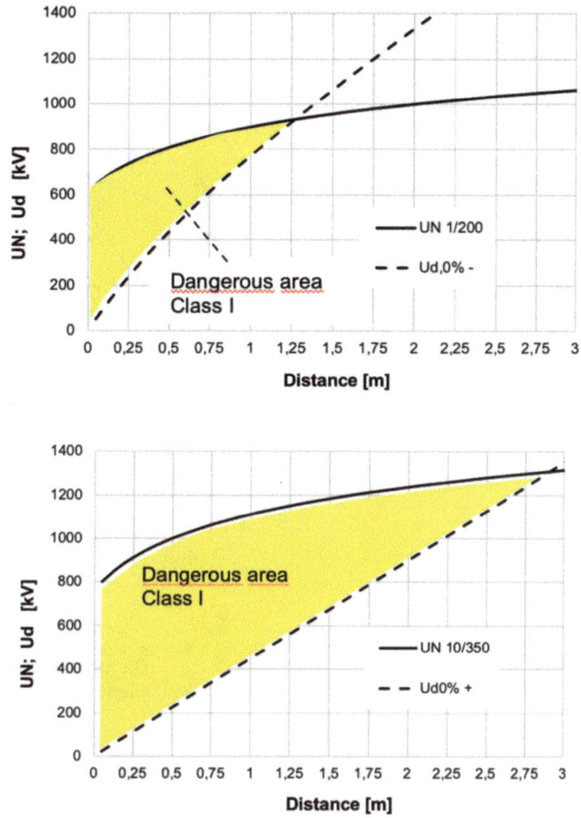

Fig. 6.28 Danger zone for proximity voltages UN and im-pulse withstand voltages in rain $U_{d,0\%}$ for the three lightning currents for values according to Table 3.1, lightning protection class I. Specific earth resistance ρ_E = 250 Ohmm. Distance of the person to the down conductor according to Fig. 6.18

6.7.2 Galvanic Coupling U_E Without a Person in Contact with Down Conductor in Case of Potential Control

The calculated arrangement corresponds to Fig. 6.6, where the person does not touch the discharge line, but the hand is a few cm away from the discharge line with the arm extended. Figure 6.34 shows two field images at different times.

- At time t = 0.5 µs the maximum current slope is present according to Fig. 6.34. Since the person does not touch the down conductor, he is subject to the influence of the lower grid of the potential control, which is at high potential. In this field image, a pronounced (capacitive) displacement current, $i_C = C * dU/dt$, where U is assumed to be proportional to the current i, can be seen through the vectors of the current density, which is directed from the surface of the person to the Ferner sheath. Therefore, the current directions in both legs are the same and the displacement current is z-directed. The corresponding current flow is shown in Fig. 6.37 and shows the one peak of about 3.5 A. This effect will be smaller for

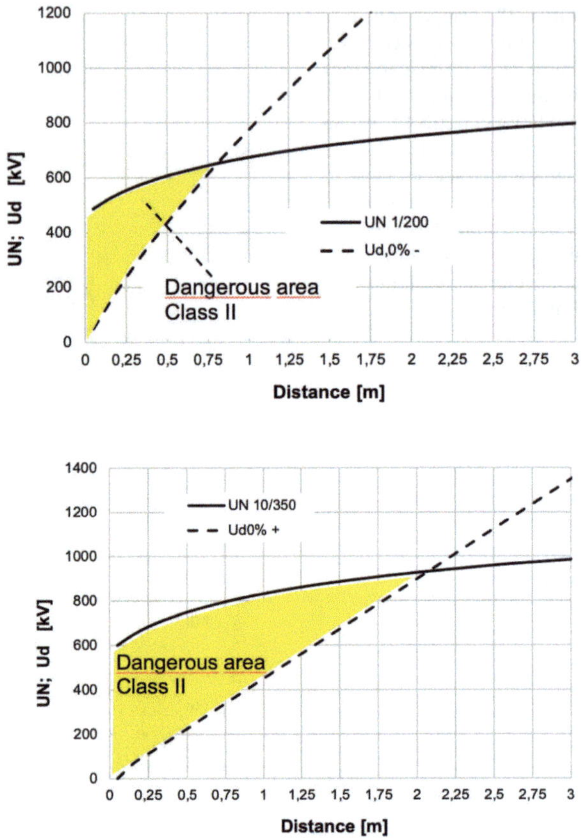

Fig. 6.29 Danger zone for proximity voltages UN and im-pulse withstand voltages in rain $U_{d,0\%}$ for the three lightning currents for values according to Table 3.1, lightning protection class II. Specific earth resistance ρ_E = 250 Ohmm. Distance of the person to the down conductor according to Fig. 6.18

the 1/200 and 10/350 flash currents because of the lower current slope and will not be analysed further here.

- At time t = 1 μs, the peak value of the exciting current and dU/dt is zero and the displacement current disappears. The coupling now becomes galvanic with the current directions in both legs differing. In Fig. 6.37, at this time, the current in the left and right legs is characterized by opposite polarity and corresponds to the vector field at time t = 1 μs. This current can be interpreted as a step voltage effect. The current is then only <0.5 A and the effect is negligible. This effect will be larger for the lightning currents 1/200 and 10/350 because of the larger current peaks and will not be analysed further here because it is negligible compared to inductive coupling.

The potential curve in Fig. 6.35 differs from that in the case with contact of the person in Fig. 6.12 because of the potential control. The person therefore has only a minor influence on the potential curve if potential control is present. The potential along a line through the arm of the person shown in Fig. 6.36. The voltage of an isolated down conductor, which is decisive for dimensioning, must therefore be

Fig. 6.30 Danger zone for proximity voltages UN and im-pulse withstand voltages in rain $U_{d,0\%}$ for the three lightning currents for values according to Table 3.1, lightning protection class III/IV. Specific earth resistance ρE = 250 Ohmm. Distance of the person to the down conductor according to Fig. 6.18

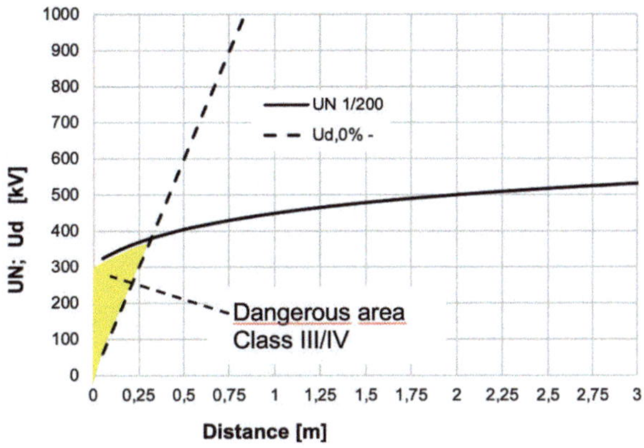

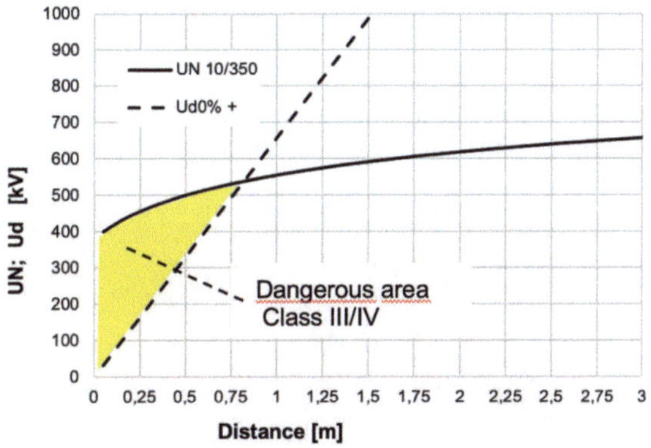

determined in the case without contact with the person in Fig. 6.36. In this case, the isolated down conductor must be dimensioned for a withstand voltage of 6 kV. With potential control, the withstand voltage of an insulated down conductor is therefore insignificant when only the galvanic coupling is considered, since the inductive coupling is predominant, see Sect. 6.3.

The potential control is of no importance with respect to the inductive coupling if the person stands on an equipotential surface, cf. Table 6.7b). Because of the large values of W and Q according to Table 6.6, no further calculation is necessary (Fig. 6.38).

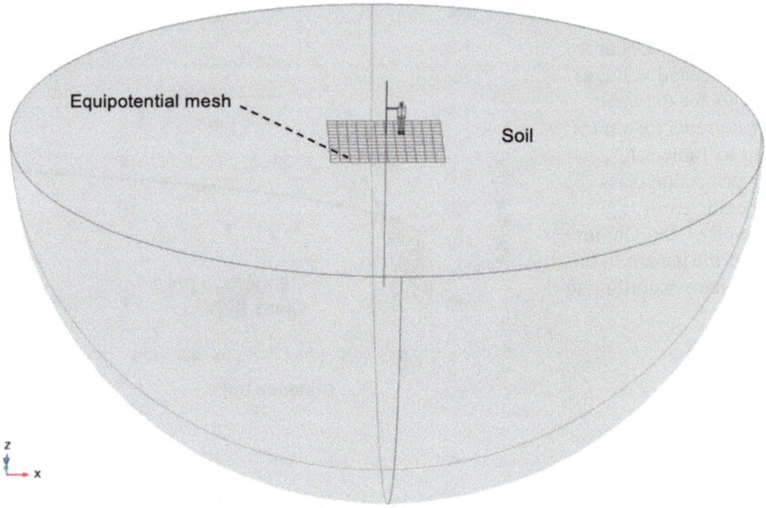

Fig. 6.31 Arrangement with potential control consisting of a square grid 6 m × 6 m with a mesh size of 0.5 m with 6 mm stainless steel bars at a depth of 0.5 m

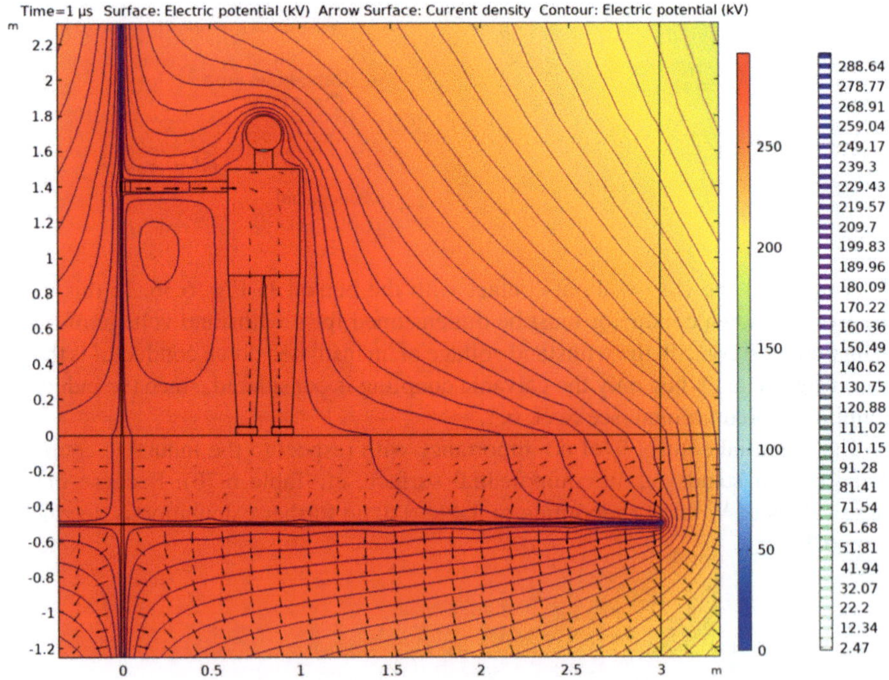

Fig. 6.32 Field image of the arrangement in Fig. 6.31 with equipotential bonding at the time of the current peak at t = 1 μs with equipotential surface and lines as well as vectors of current density. Negative following flash 0.25/100 with a peak value of 24.7 kA according to Table 3.1. Soil ρ_E = 250 Ohmm. Current density vectors in logarithmic representation

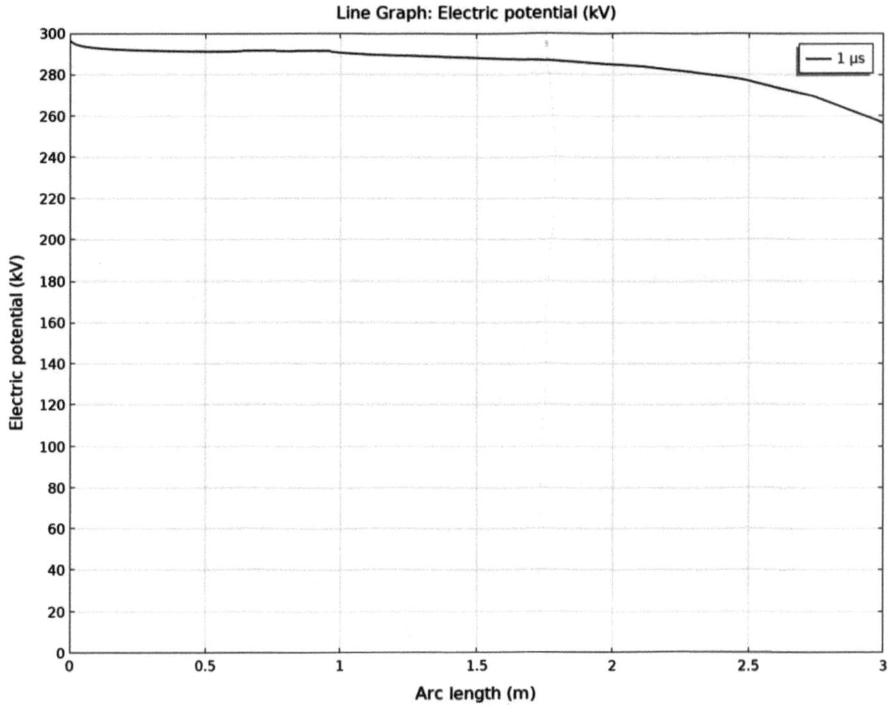

Fig. 6.33 Potential curve on the earth's surface along line B in Fig. 6.9 at the time of the current peak t = 1 μs for 0.25/100. Arrangement with potential equalization

Table 6.6 Results for site isolation. Type A earth electrode; lightning protection class I. Energy values for the other lightning protection classes are shown in parentheses. Exceeded limit values are shown in bold

	I_B	V_{Hand}	V_{Foot}	U_P	I_K	W	Q
	kA	kV	kV	kV	A	Ws	mAs
+10/350	52	622	613	9	9.5	**23.2 (I)** **13.05 (II)** **5.8 (III–IV)**	**4.7**
−1/200	42	503	495.7	7.5	7.7	**8.5 (I)** **4.8 (II)** **2.13 (III–IV)**	**2.2**
−0.25/100	24.7	296	291.5	4.5	4.5	**1.5 (I)** 0.84 (II) 0.375 (III–IV)	**0.6**

I_B Peak value of the injected lightning current according to Table 3.1
I_K Peak value of the body current at the time of the current peak
V_{Hand} Potential at the point of contact of the hand against distant earth
V_{Foot} Potential at the foot against distant earth
U_P Voltage between hand and foot

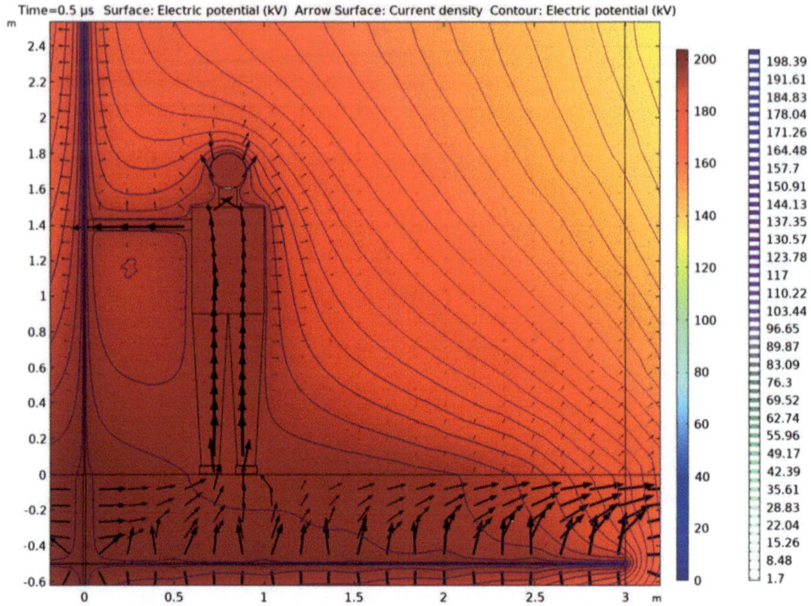

Field plot at t = 0,5 µs

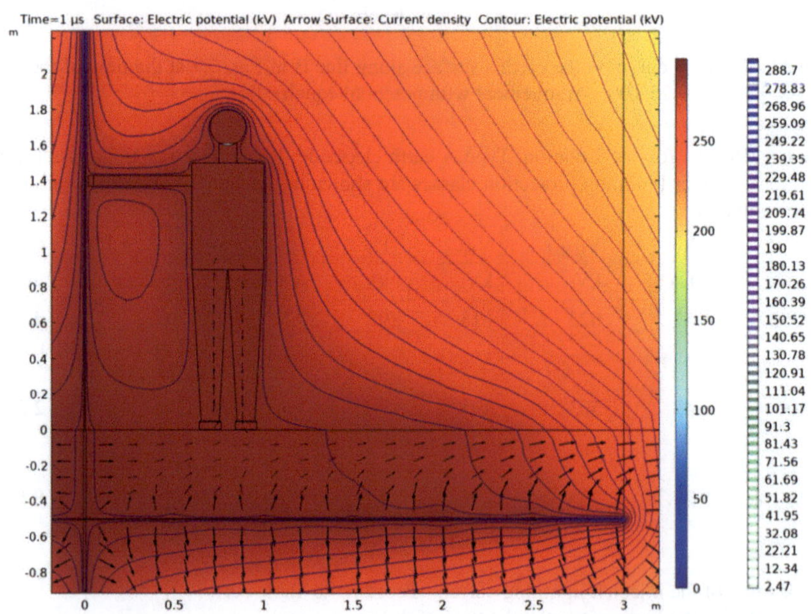

Field plot at t = 1 µs

Fig. 6.34 Field images for the case when the person does not touch the down conductor. Field image at the time of the current peak at t = 0.5 µs and 1 µs with equipotential surface and lines as well as vectors of current density. Negative following flash 0.25/100 with a peak value of 24.7 kA according to Table 3.1. Soil $\rho_E = 250$ Ohmm. Current density vectors in logarithmic representation

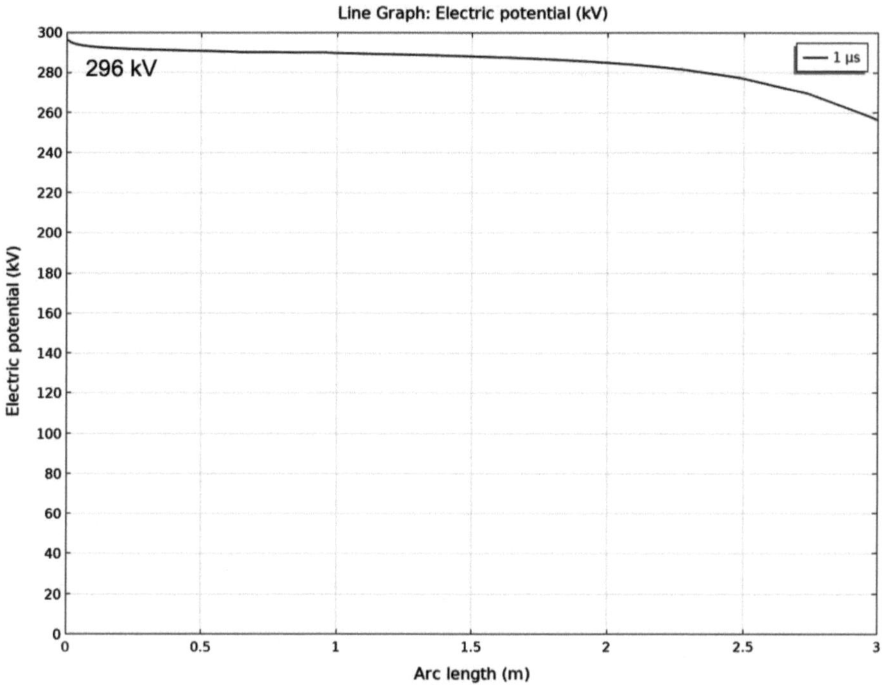

Fig. 6.35 Potential curve on the earth's surface along line B in Fig. 6.9 at the time of the current peak t = 1 µs for 0.25/100

6.8 Results

6.8.1 Person in Contact with a Down Conductor in a Distance of 0.8 m

The greatest danger is the galvanic coupling U_E. Table 6.7a) shows for all lightning current forms that a person dies when approaching or touching the discharge. Also, the attempt to zero the high voltage U_E by a perfectly conducting plate (perfect equipotential bonding) under the person leads to the values in Table 6.7b). There it can be seen that the galvanic coupling of the down conductor and the inductive coupling, despite perfect equipotential bonding, lead to much too high values of W and Q for the lightning currents 0.25 and 1/200, so that a person dies when touching the down conductor. However, these values apply to the analytical lightning current forms according to EN 62305-1 (Heidler functions), for which the current slope is 1.38 times greater than the values specified in EN 62305-1, Table 3.1, see Fig. 17.8, and Sect. 15.2. Table 6.8 therefore shows the current slope for comparison for the values specified in EN 62305-1, Table 3.1. The values for W and Q reach harmless values in lightning protection class III/IV! The voltage waveforms on a person touching

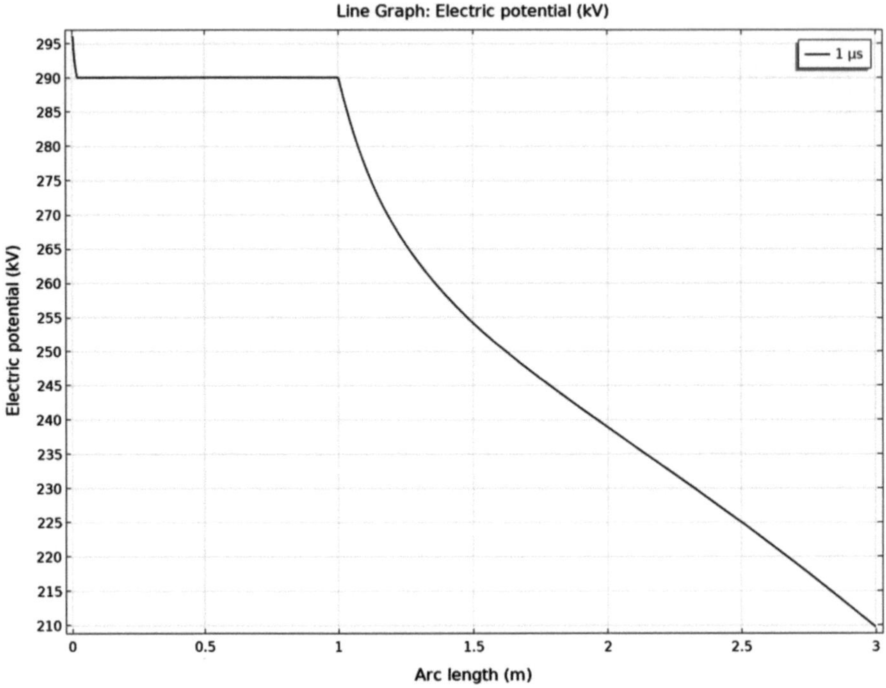

Fig. 6.36 Potential curve along line B in Fig. 6.13 at the time of the current peak t = 1 μs for 0.25/100

the down conductor are shown in Fig. 6.39 for inductive coupling and correspond to the time derivative of the impressed lightning current. Figure 6.40 shows that the resulting voltage of all couplings leads to the same peak value as for purely galvanic coupling. Figure 6.41 shows the voltage shapes that occur with galvanic coupling that are proportional to the impressed lightning current. For negative cloud-to-ground lightning, multiple lightning discharges can occur. In this case, the energy W and the charge Q have a cumulative effect, i.e. the values for 1/200 and 0.25/100 in Table 6.7 must be added together. Even in the case where a person does not touch the discharge directly, there is danger from the proximity voltage U_N, the cause of which is the galvanic coupling. In this case, the positive lightning is the greatest danger. This case is the main cause of fatalities, see Chap. 14, when persons seek shelter e.g. under a tree. The necessary safety distances are in the range of a few meters and can be even greater if the ground is dry.

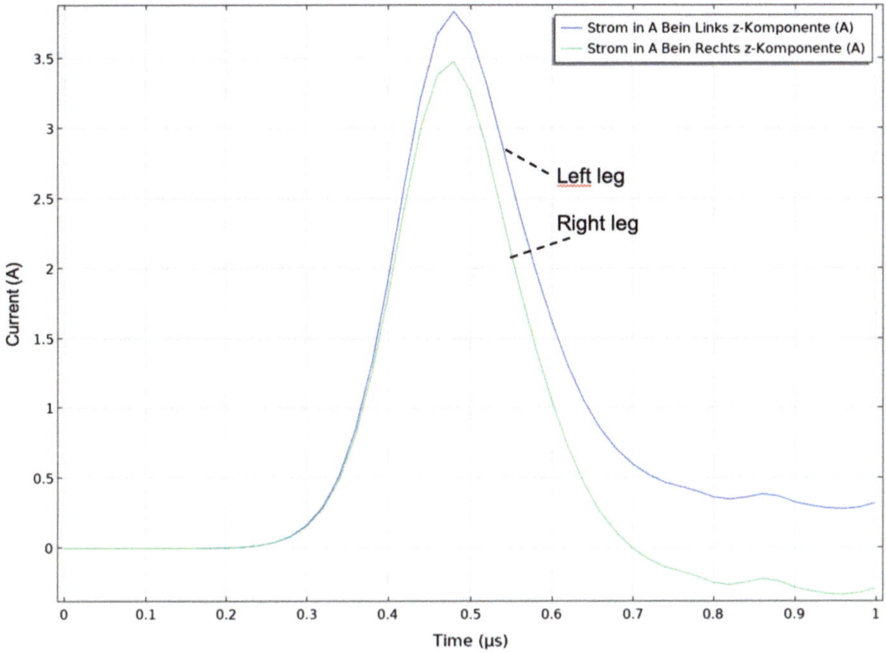

Fig. 6.37 Current waveform of the z-component (ec.Jz) in both legs for 0.25/100 when the person does not touch the down conductor

6.8.2 Galvanic Coupling with a Person in Contact with Down Conductor in a Distance of 0.4 m

In case of a distance of a person of 0.4 m to the down conductor, the load of the person decreases in all lightning protection classes. The potential at the earth's surface shows a smaller voltage drop along the person for the case with 0.4 m distance than in the case of a distance of 0.8 m in Fig. 6.10. Accordingly, the energy converted in the person is also smaller. Table 6.9 shows the values for W and Q for the case of the distance of 0.4 m. In lightning protection class I to III/IV, however, the values W and Q remain far above the limit value (Fig. 6.43).

6.8.3 Results for Inductive Coupling Depending on the Distance of a Person to the Down Conductor

The effect of inductive coupling is negligible in the arrangement according to Fig. 6.6 compared to galvanic coupling. However, with regard to the further investigations

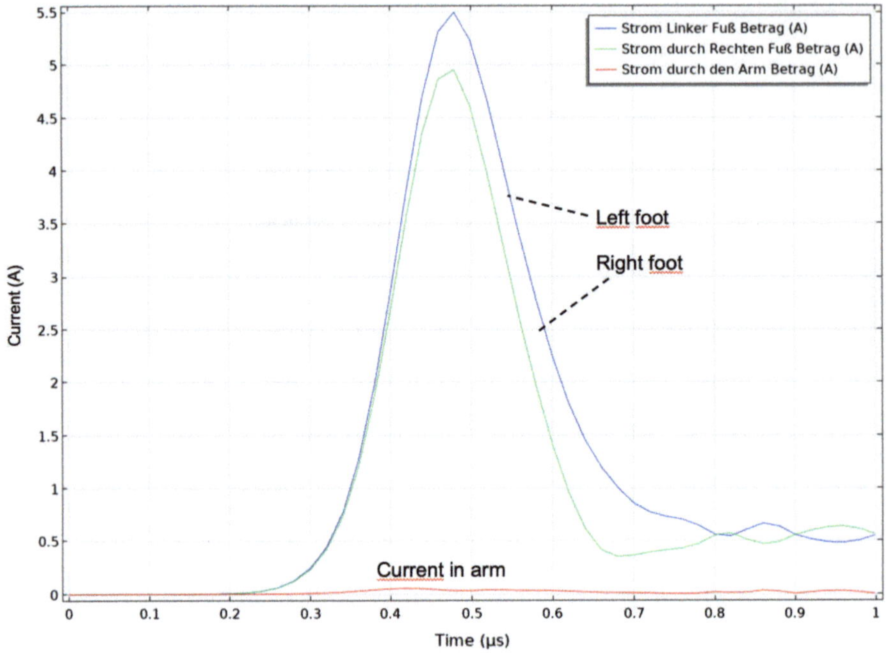

Fig. 6.38 Current waveform (magnitude) in both legs and arm for 0.25/100 when the person does not touch the down conductor. Inductive coupling U_M with potential control

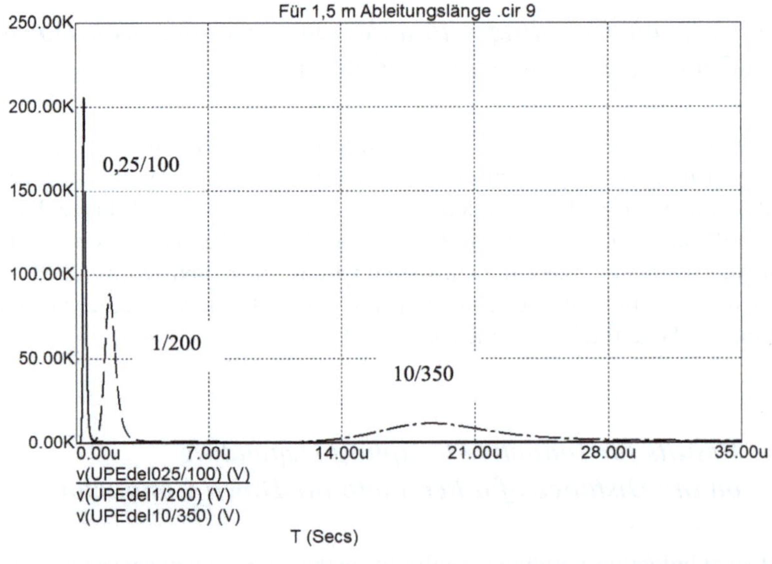

Fig. 6.39 Forms of the voltage occurring on a person at the three lightning currents according to IEC 62305-1. (Galvanic coupling of the discharge and inductive coupling.) Calculation for values in Table 3.1, earth electrode type A, stainless steel 8 mm with Heidler formulas [4]

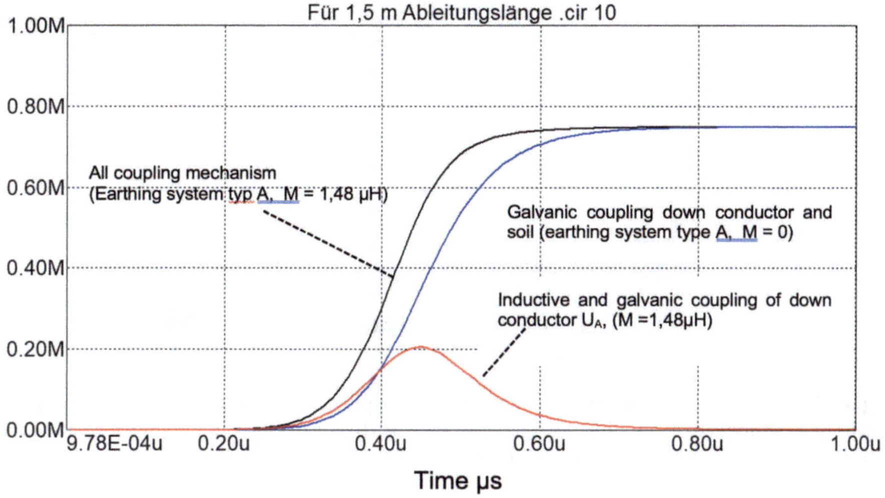

Fig. 6.40 Example comparison of individual voltage components for lightning current 0.25/100. Calculation for values in Table 3.1, Erder type A, stainless steel 8 mm using Heidler formulas [4][1]

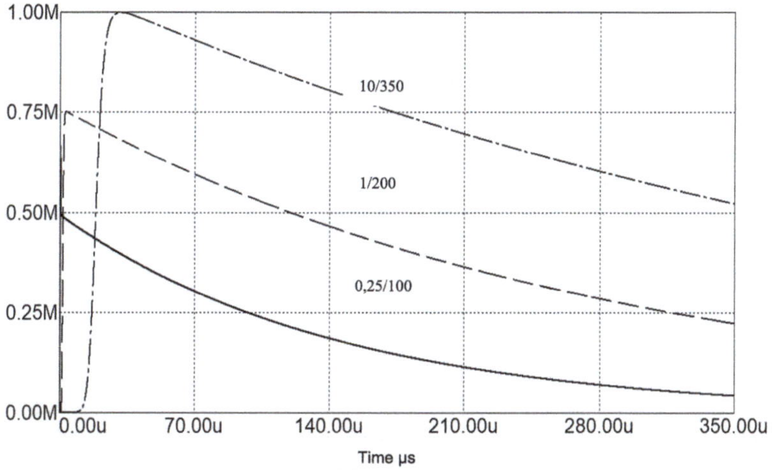

Fig. 6.41 Shape of the voltage occurring at a person at the three lightning currents according to EN 62305-1, valid for all couplings. Calculation for values in Table 3.1, earth electrode type A, stainless steel 8 mm with Heidler formulas [4][2]

[1] See 4.

[2] The steepness of the current in case of Heidler functions is about 40% higher compared to the fixed values [30].

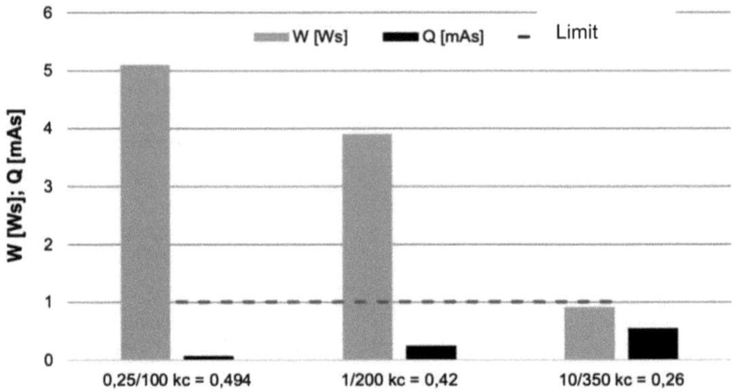

Fig. 6.42 Energy and charge only for the galvanic coupling in the down conductor (negligibly small) and the inductive coupling according to Table 6.7b) for the three lightning currents (earth electrode type A) with kc values according to Table 3.1. Calculation with Heidler formulas [4] in Microcap

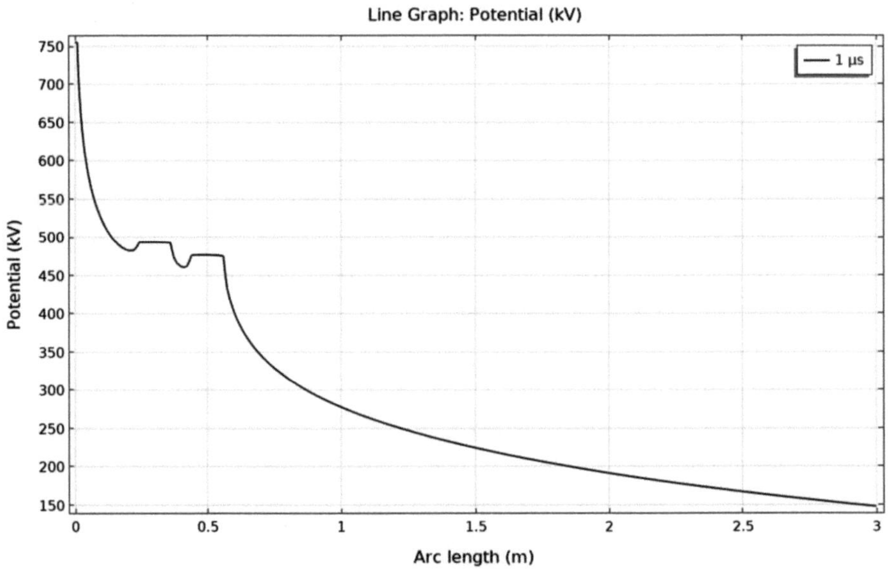

Fig. 6.43 Potential on the earth's surface along line B in Fig. 6.9 at the time of the current peak t = 1 μs for 0.25/100 at a **distance of the person of 0.4 m from the down conductor**

with site insulation and equipotential bonding as well as with insulated down conductors, the inductive coupling is in the foreground. Therefore, a more detailed assessment is made at this point. In particular, it should also be noted that people can touch the down-conductor even at distances smaller than 0.8 m.

Table 6.7 W, Q, voltage peak U_P, and current i_K in a person touching a down conductor, calculated with Comsol-Multiphysics and in the equivalent circuit according to Fig. 6.17 for values in Table 3.1 for earth electrode type A, stainless steel 8 mm. Person touching the down conductor at a height of H = 1.4 m above ground at a distance of 0.8 m from the center of the down conductor, see Fig. 1.1. Exceeded limit values are shown in bold. Values apply to lightning protection class I and the analytical lightning current shapes (Heidler functions) specified in IEC 62305-1. Values for lightning protection classes II and III/IV in footnotes (1) and (2)

				(a) *Galvanic coupling (down conductor and soil) (earthing system Typ A, M=0)*			
	kc	I	U_P Hand-Foot	$V_P^{(1)}$ Potential on hand/down conductor	i_K	W	Q
		kA	kV	kV	**A**	**kWs**	mAs
10/350	0.26	52	767	1592	**767**	**150.9**	**388**
1/200	0.42	42	621	1286	**621**	**55.7**	**178**
0.25/100	0.494	24.7	363	755.5	**363**	**9.5**	**52**
				(b) *Inductive coupling and galvanic coupling of down conductor U_A ($M = 1.48\ \mu H$)*			
			U_P Hand -Foot	$V_P^{(2)}$ Potential on hand/down conductor	i_K	W	Q
	kc	I	kV	kV	**A**	**Ws**	mAs
10/350	0.26	52	11.4	11.4	**11.4**	0.91	0.55
1/200	0.42	42	88.7	88.7	**88.7**	**3.9** (5)	0.25
0.25/100	0.494	24.7	205	205	**205**	**5.1**(6)	0.076
				(c) *All coupling mechanism (earthing system Typ A, $M = 1.48\ \mu H$)*			
			$U_P^{(2)}$ Hand-Foot	$V_P^{(3)}$ Potential on hand/down conductor	$i_K^{(4)}$	W	Q
	kc	I	kV	kV		**kWs**	mAs
10/350	0.26	52	767	1592		**150.9**	**388.5**
1/200	0.42	42	621	1286		**55.7**	**178.25**
0.25/100	0.494	24.7	363	755.5		**9.5**	**52.07**

kc Coefficient of current distribution in conductors
I Effective peak value of lightning current with kc in lightning protection class I
i_K Peak value of lightning current in the body of the touching person
U_P Voltage applied between the hand and the bottom of the foot
V_P Potential at the hand
(1) The voltage waveform corresponds approximately to the applied pulse waveform of the current shown in Fig. 6.36
(2) Peak values in Fig. 6.30 are shifted at 0.25/100 because of the induced voltage, but show equal peak values
(3) Peak values in Fig. 6.30 are shifted in time, addition is not reasonable
(4) See (3)
(5) For BSK II: 2.2 Ws; for Class III/IV: 1.23 Ws
(6) For BSK II: 2.86 Ws; for Class III/IV: 1.61 Ws

Table 6.8 Supplement to Table 6.7. Inductive coupling and galvanic coupling of the down conductor for the values of the maximum specified lightning current parameters specified in accordance with EN 62305-1, Table 6.1, see Fig. 17.8 for lightning protection class I. Values for lightning protection class II and III/IV in footnotes (1) and (2)

				(a) *Inductive coupling and galvanic coupling of the down conductor U_A (**$M = 1.48\ \mu H$**) for values as per IEC 62305-1, Table 6.1*			
			U_P Hand-foot	V_P(2) Potential on hand/down conductor	i_K	W	Q
	kc	I	kV	kV	**A**	**Ws**	mAs
10/350	0.26	52	8.26	8.26	**8.26**	0.47	0.72
1/200	0.42	42	64.3	64.3	**64.3**	**2.04**(1)	0.218
0.25/100	0.494	24.7	149	149	**149**	**2.68**(2)	0.055

(1) For Class II: 1.14 Ws; for Class III/IV: 0.51 Ws
(2) For Class II: 1.5 Ws; for Class III/IV: 0.67 Ws

Table 6.9 W, Q, peak voltage UP, and current i_K in a person touching a down conductor at a distance of 0.4 m, calculated with Comsol-Multiphysics and in the equivalent circuit figure according to Fig. 6.17 for values in Table 3.1 for earth electrode type A, stainless steel 8 mm. Person touches the down conductor at a height of H= 1.4 m above ground at a distance of 0.4 m from the centre of the down conductor, see Fig. 2.1. Exceeded limit values are shown in bold. Values apply to lightning protection class I and the analytical lightning current forms (Heidler functions) specified in IEC 62305-1

				Galvanic coupling of the down conductor and soil (earthing system type A, M=0) Class I, distance 0.4 m			
	kc	I	U_P Hand foot	V_P(1) Potential on hand/down conductor	i_K	W	Q
		kA	kV	kV	**A**	**kWs**	mAs
10/350	0.26	52	570 (767)	1592	**578** (767)	**85.8** **(150.9)**	**292** **(388)**
1/200	0.42	42	455 (621)	1285	**455** (621)(1)	**29.9** (55.7)	**130** (178)
0.25/100	0.494	24.7	275 (363)	755	**275** (363)(1)	**5.47** (9.5)	**39.4** (52)

(1) Values for 0.8 m distance in brackets

6.8.3.1 Inductive Coupling with a Person in Contact with the Down Conductor

In the previous calculations, a fixed distance of 0.8 m between the person and the discharge at a height of 1.4 m was assumed. This assumption is for a statistically extremely rare case. It is easy to see that a person does not touch a down conductor

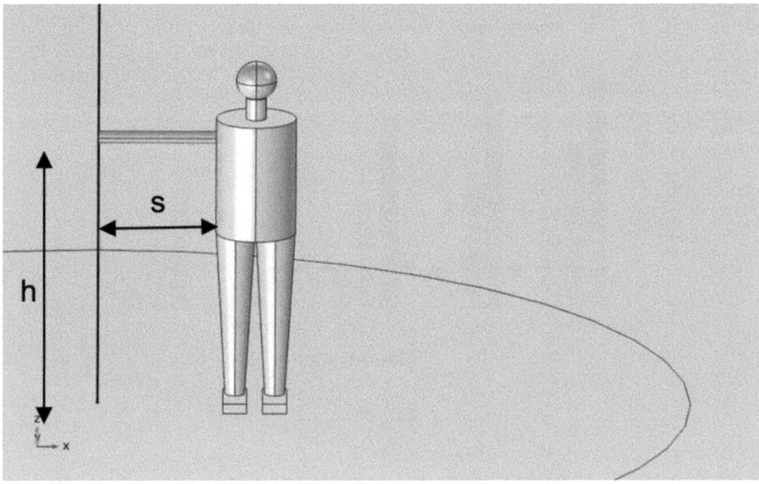

Fig. 6.44 Person in contact with down conductor in a distance s and height h

with an outstretched arm during an approaching thunderstorm or during a thunderstorm, see also Chaps. 14–16. On the other hand, it is more likely that a person approaches the down conductor at a distance s according to Fig. 6.44 and touches the down conductor with an outstretched, bent arm or shoulder. For this case, the energy W transferred in the person is shown as a function of the distance s in Fig. 6.45. It follows from this that there is no danger up to a distance of 60 cm from a negative first flash (without a subsequent flash) in lightning protection class III/IV. It also follows that there is a hazard in the case of cumulative exposure to the negative first flash and a subsequent negative second flash. It should be noted that only about 50% of all negative first flashes contain a negative subsequent flash. If the person touches with the lower extremities, the height h is reduced and the energy W can be converted quadratically with the reduced height.

Touching the down conductor is therefore the greatest danger of the case of a negative first stroke with a negative subsequent stroke.

6.8.3.2 Inductive Coupling Without a Person in Contact with the Down Conductor with Risk of a Flash Over

If a person approaches the arrester, a flashover can occur between the arrester and the person, provided the person is standing within the danger zone according to Sect. 17.2.2. The worst case here is the negative first lightning and in lightning protection class I. In lightning protection class I, flashover can therefore only occur at a distance $s < 9$ cm and, according to Fig. 17.9, the cumulative value would exceed the limit value of 1 Ws. In lightning protection class II/IV, according to Fig. 17.11, there is a danger zone for the negative first flash at a distance of 4 cm. This no longer

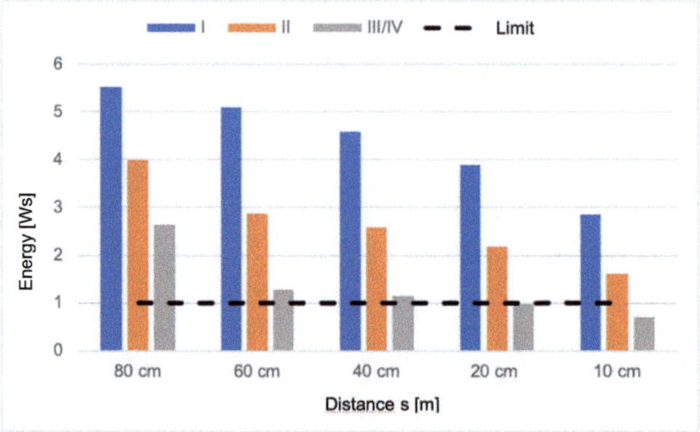

0,25/100

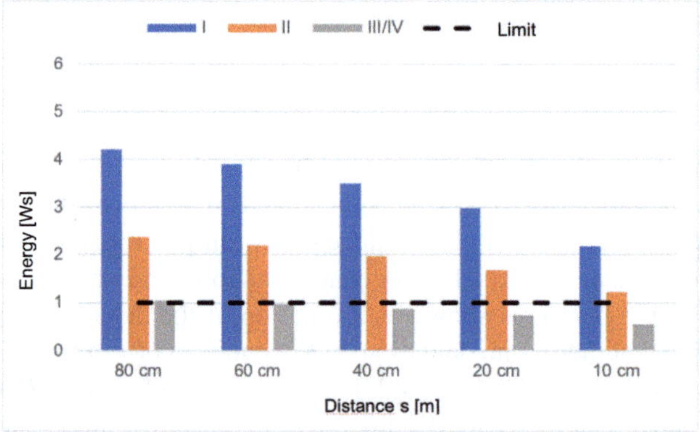

1/200

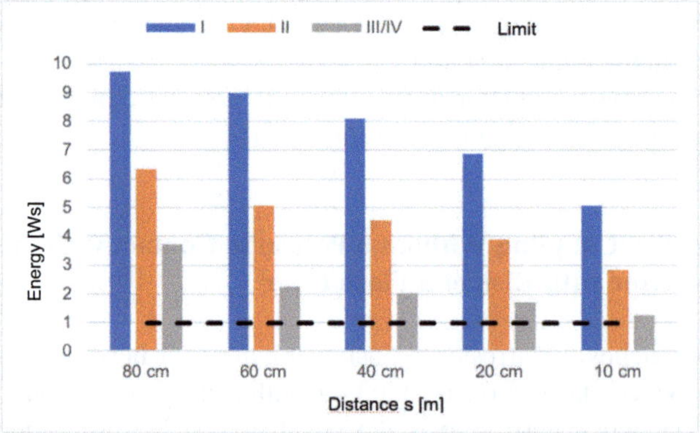

Cumulative values 1/200 +0,25/100

Fig. 6.45 Energy W dissipated in one person **when touching a down conductor** at a height of 1.4 m at variable distance s according to Fig. 6.44

represents a hazard even with the cumulative energy. A person who approaches the down conductor to a distance of 4 cm during a thunderstorm is no longer at risk in lightning protection class III/IV, even if a flashover occurs at a distance of 4 cm. Further measures such as insulated down conductors are therefore required if full protection is to be guaranteed in lightning protection class I and II.

Chapter 7
Effect of Site Insulation with Asphalt According to IEC 62305-3

To limit the contact stress, [12] suggests site insulation with a 5 cm thick layer of asphalt or a 15 cm thick layer of gravel. In [12], no distinction is made between dry and wet conditions. Likewise, [12] does not describe the influence of a water layer on the asphalt due to rain/heavy rain during or before a thunderstorm. A calculation with dry asphalt is therefore not realistic and instead the calculations are carried out with wet asphalt with a specific resistance of 10,000 Ωm. However, conclusions for the behaviour of dry asphalt can be derived from the results. Another calculation is carried out with a 2 cm water layer on an asphalt layer. The calculation of the investigated variants is carried out numerically according to the information in Sect. 6.2. According to Fig. 7.1, an asphalt layer with a thickness of 5 cm and a diameter of 6 m is placed on the ground surface flush with the ground surface. The asphalt has a resistivity of $\rho_E = 10{,}000$ Ωm and a relative permittivity of $\varepsilon_r = 3$, which according to Table 5.1 corresponds to asphalt in the wet state.

7.1 Galvanic Coupling U_E with Contact of a Person with the Down Conductor

According to Fig. 7.2, this results in a current flow through the body, the effect of which is evaluated and shown with further values in Table 7.1. In particular, a capacitive current component acts in the case of the negative subsequent stroke, which is proportional to the derivative of the impressed current. However, this energy and charge component is negligible compared to the galvanic coupling. The galvanic coupling results in a massive exceeding of the limit values, so that a person dies. The course of the earth surface potential along the surface of the asphalt layer is shown in Fig. 7.6 and provides an insight into the physical conditions. A voltage of 85 kV drops across the person and a voltage of 678 kV results between the foot and the distant earth, Table 7.1. Another finding from the simulation is the high voltage at the foot of the person. The evaluation of the electric field strength in Figs. 7.3 and

J. Meppelink, *The Lightning Rod as a Danger*,
https://doi.org/10.1007/978-3-031-30434-7_7

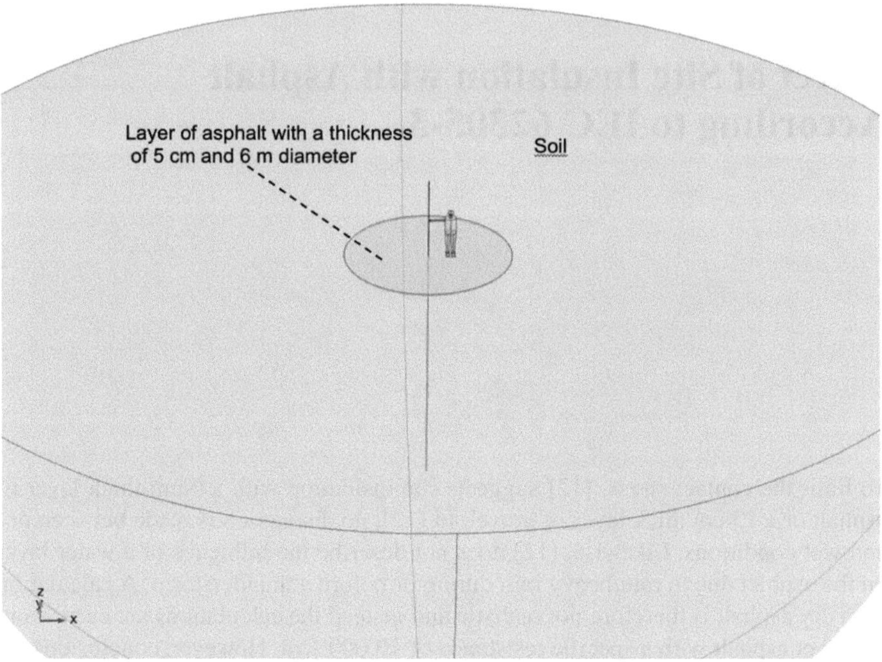

Fig. 7.1 Asphalt layer with a thickness of 5 cm and 6 m diameter

Table 7.1 Results for site insulation with wet asphalt with ρ = 10,000 Ωm. Type A earth electrode; Class I. Exceeded limit values are shown in bold

	I_B	V_{Hand}	V_{Foot}	U_P	I_K	W	Q
	kA	kV	kV	kV	A	kWs	mAs
+10/350	52	1610	1440	170	178	**8.128**	**87.9**
−1/200	42	1299	1163	136	144	**2.996**	**41**
−0.25/100	24.7	763	678	85	85	**0.520**	**12.2**

I_B—Peak value of the injected lightning current according to Table 3.1
I_K—Peak value of the body current at the time of the current peak
V_{Hand}—Potential at the point of contact of the hand against distant earth
V_{Foot}—Potential at the foot against distant earth
U_P—Voltage between hand and foot

7.4 shows a field component perpendicular to the surface of the asphalt, which is known to lead to the formation of surface discharges, which are described in more detail in Sect. 20.1. The normal component of the electric field strength is shown in Fig. 7.5. The value of 25 kV/cm for the insertion field strength of electric discharges in air is clearly exceeded in the foot area. The corona inception voltage of a surface discharge voltage is exceeded according to Fig. 20.4, so that a corona occurs in the foot area. The leader inception voltage of a surface discharge in Fig. 20.5 is in the

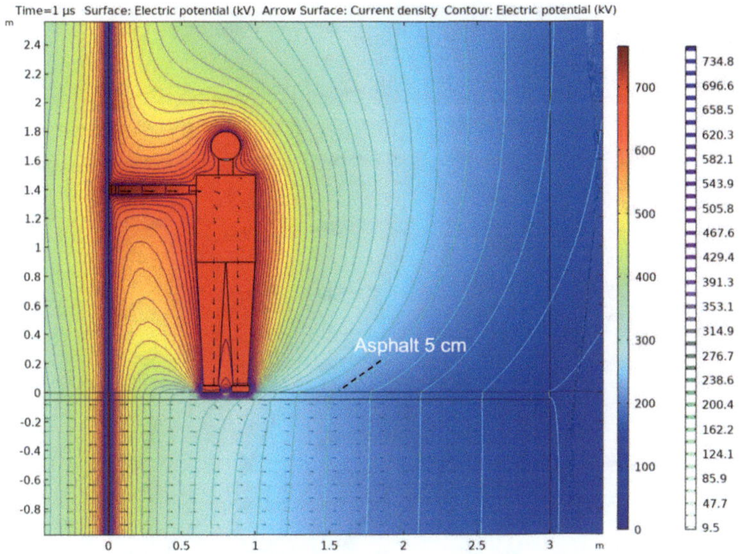

Fig. 7.2 Field plot at the time of the current peak at t = 1 μs with equipotential surfaces and lines as well as vectors of the current density. Negative subsequent stroke 0.25/100 with a peak value of 24.7 kA according to Table 3.1. ρ_E = 250 Ωm. ε_r = 3. Asphalt ρ = 10,000 Ωm, ε_r = 3. Current density vectors in logarithmic representation

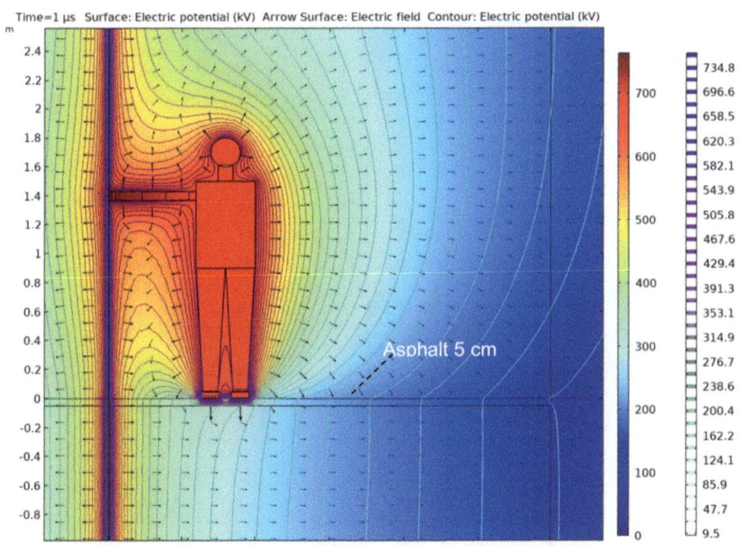

Fig. 7.3 Electric field at the time of the current peak at t = 1 μs with equipotential surfaces and lines as well as vectors of the electric field strength. Negative subsequent stroke 0.25/100 with a peak value of 24.7 kA according to Table 3.1. Soil ρ_E = 250 Ωm, ε_r = 3. Asphalt ρ = 10,000 Ωm, ε_r = 3. Field strength vectors in logarithmic representation

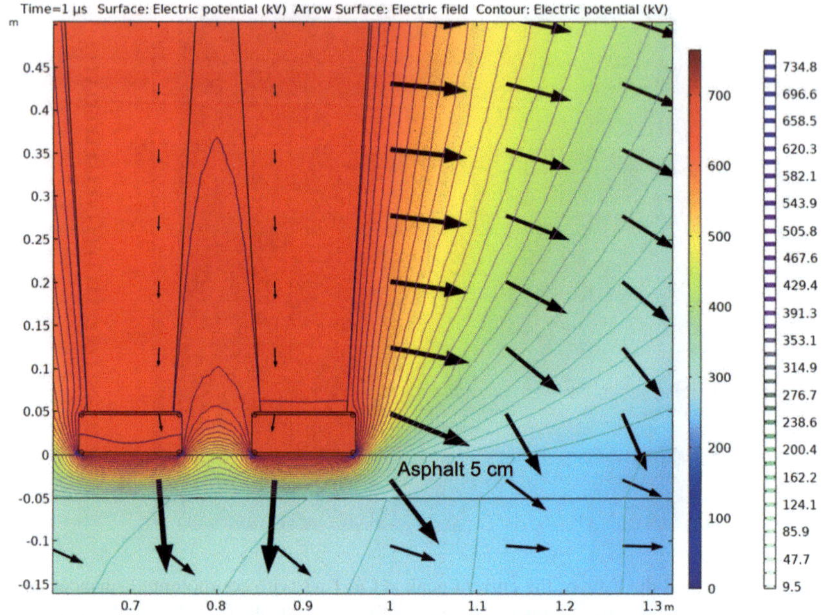

Fig. 7.4 Enlarged section from Fig. 7.3

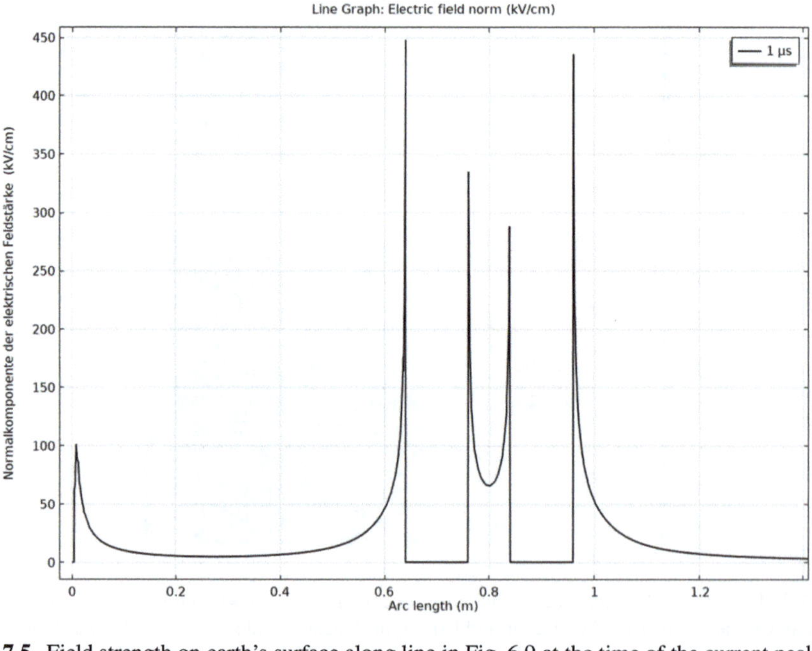

Fig. 7.5 Field strength on earth's surface along line in Fig. 6.9 at the time of the current peak t = 1 µs for 0.25/100

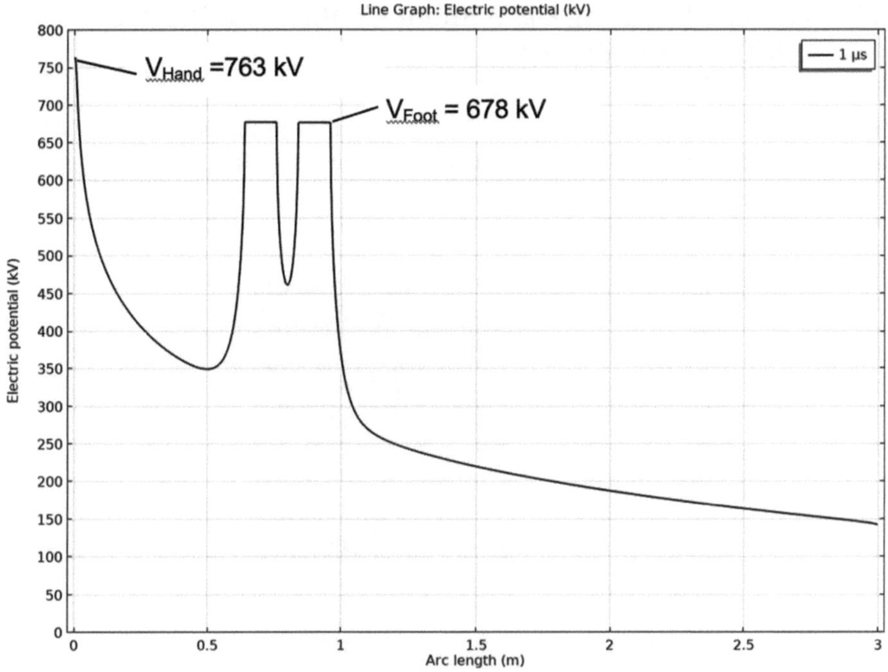

Fig. 7.6 Potential on earth's surface along line in Fig. 6.9 at the time of the current peak t = 1 μs for 0.25/100

range of 120 kV for 5 cm thick asphalt while the applied voltage is 678 kV, resulting in surface discharges. The field figures for the lightning currents 1/200 and 10/350 are similar. With dry asphalt, the resistivity would increase significantly, so the current through the body would decrease further. However, the high voltage at the foot of the person would continue to cause surface discharges at the foot. This would bridge the insulating distance between the arrester and the foot and a person would then die from the effects of the surface discharge. A recalculation for lightning protection class III shows no improvement; here, too, a person would die from excessive energy and/or surface discharge.

7.2 Inductive Coupling U_M with a Person in Contact with the Down Conductor

Inductive coupling is of secondary importance compared to the great danger of galvanic coupling and the danger of surface discharges and is therefore not calculated in detail.

7.3 Galvanic Coupling U_E Without Contact of a Person with Down Conductor

The calculated arrangement corresponds to Fig. 7.1, with the person not touching the down conductor. Figure 7.7 shows two field figures at different times (Figs. 7.8, 7.9, 7.10, 7.11 and Table 7.2).

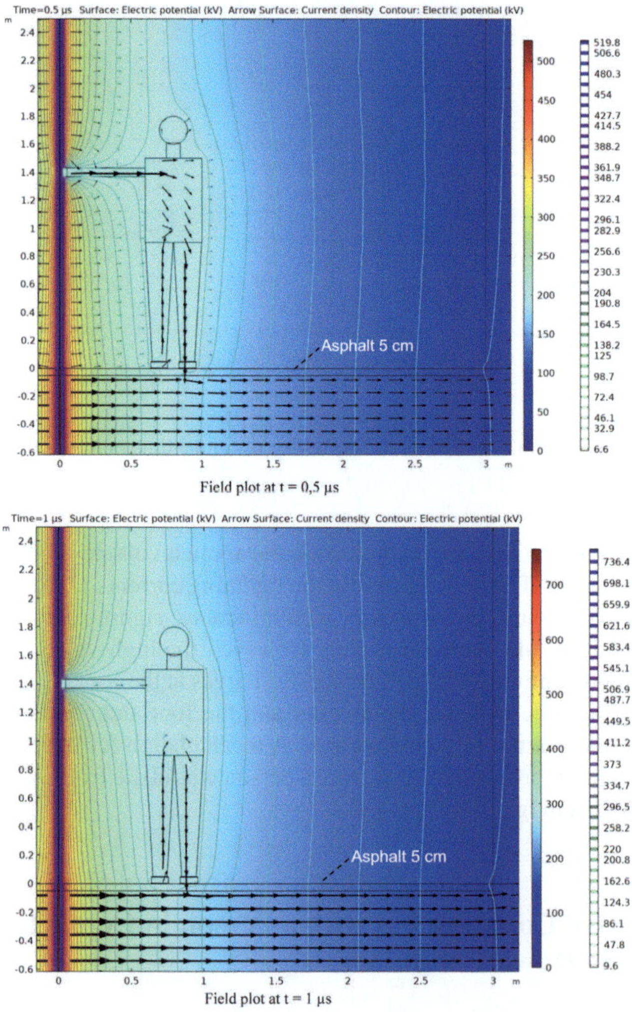

Fig. 7.7 Field for the case that the person **does not touch** the down conductor. Field at the time of the current peak at t = 0.5 μs and 1 μs with equipotential surface and lines as well as vectors of current density. Negative subsequent stroke 0.25/100 with a peak value of 24.7 kA according to Table 3.1. Soil $\rho_E = 250\ \Omega m$, $\varepsilon_r = 3$. Asphalt $\rho = 10.000\ \Omega$, $\varepsilon_r = 3$. Vectors of current density in logarithmic representation

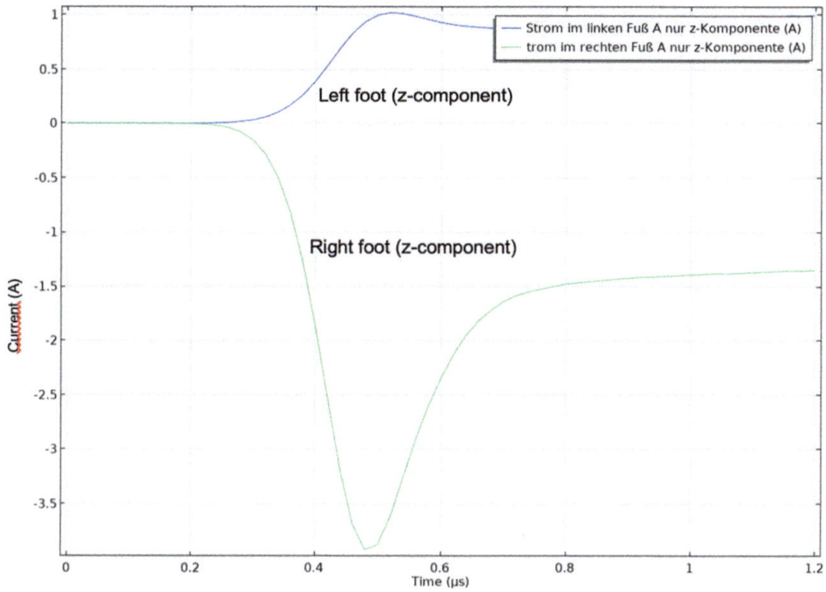

Fig. 7.8 Current (z-component) in both legs for 0.25/100. The person does not touch the down conductor

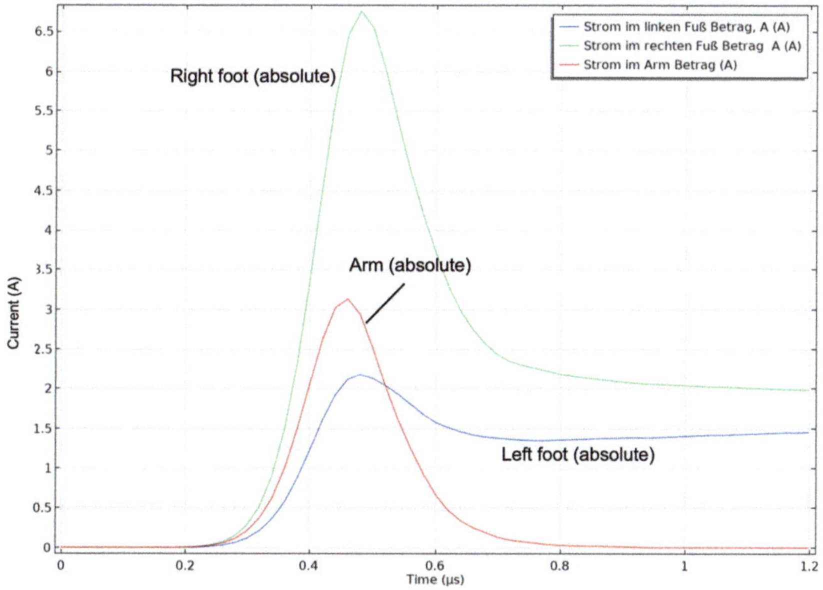

Fig. 7.9 Current (Absolute) in both legs for 0.25/100. The person does not touch the down conductor

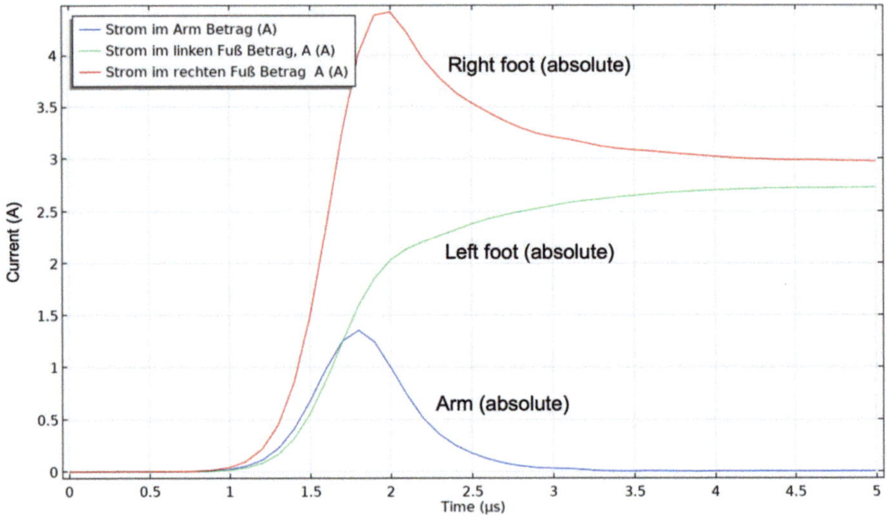

Fig. 7.10 Current (Absolute) in both legs and arm for 1/200. The person does not touch the down conductor

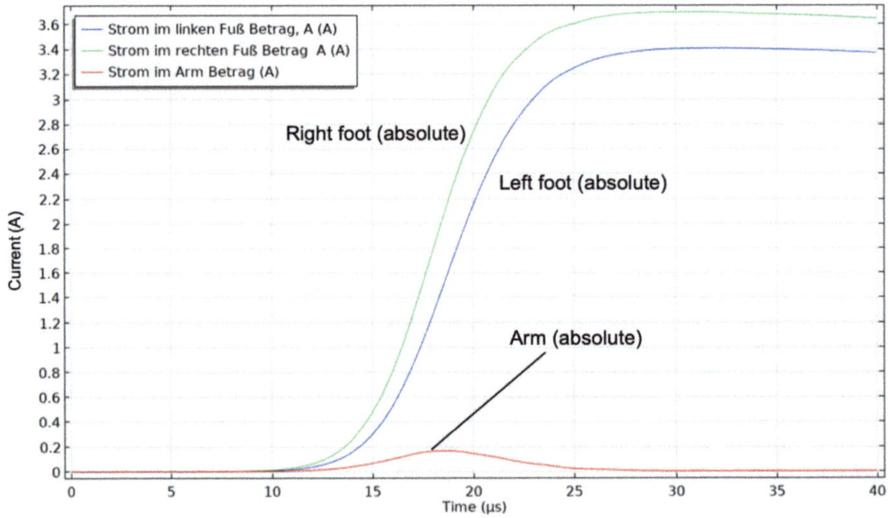

Fig. 7.11 Current (Absolute) in both legs and arm for 10/350. The person does not touch the down conductor

Table 7.2 Magnitude of current I. Earthing system type A; Class I

	I_B	I	Limit for step voltage according to [34]
	kA	A	A
10/350	52	3.7	25
1/200	42	3	No limit value available. The value for 10/350 can be applied
0.25/100	24.7	2	

I_B—Peak value of the injected lightning current according to Table 3.1
I—Peak value of the current in the body of a person

At time **t = 0.5 μs** the maximum current slope prevails. In this field figure, a pronounced displacement current can be seen through the vectors of the current density. Since the person does not touch the down conductor, he is subject to the influence of the ground below, which is at a high potential. The corresponding current flow is shown as a z-component in Fig. 7.8 and as a magnitude in Fig. 7.9 and shows a peak of about 3 A. The current is a real displacement current and proportional to di/dt. This effect will be smaller for the lightning currents 1/200 and 10/350 due to the lower current slope and will not be analysed further here.

At time **t = 1 μs** the peak value of the exciting current is reached and the coupling becomes predominantly galvanic. In Fig. 7.7, the current in the left and right leg at this time is characterised by opposite polarity and corresponds to the vector field at time t = 1 μs. The opposite current flow in the legs is the step voltage effect. The step voltage effect also occurs at a step width below the standardised step width of 1 m. The amount of the current in Fig. 7.9 is then only 1.5–2 A in the further course after reaching the current peak and is negligible in its effect. This effect is slightly larger for the lightning currents 1/200 and 10/350 because of the larger current peak values and is shown in Figs. 7.10 and 7.11 as well as in the results in Table 7.2. The step voltage effect is present but not critical.

The potential curve in Fig. 7.12 differs from that in the case with contact of the person in Fig. 7.6. The person therefore only has a minor influence on the potential curve if potential control is present. The potential along a line through the arm of the person is shown in Fig. 7.13. A voltage of 481 kV drops between the discharge and the person, which leads to a flashover and causes the same lethal effects as in the case of contact, shown in Sect. 7.1.

The voltage of an insulated down conductor that is decisive for evaluation must therefore be determined in the case without contact with the person in Fig. 7.13. In this case, an insulated down conductor must be designed for a withstand voltage of 481 kV in the case of a negative sequence lightning due to the galvanic coupling. For the other types of lightning currents, the withstand impulse voltages are listed in Table 7.2.

7.4 Electric Coupling Due to Proximity U_N

As already described in Sect. 6.4, a flashover to a person (side flash) also occurs in the case of site insulation with wet asphalt if the person is at a distance from the down-conductor. The flashover between the arrester and the hand is then given by the

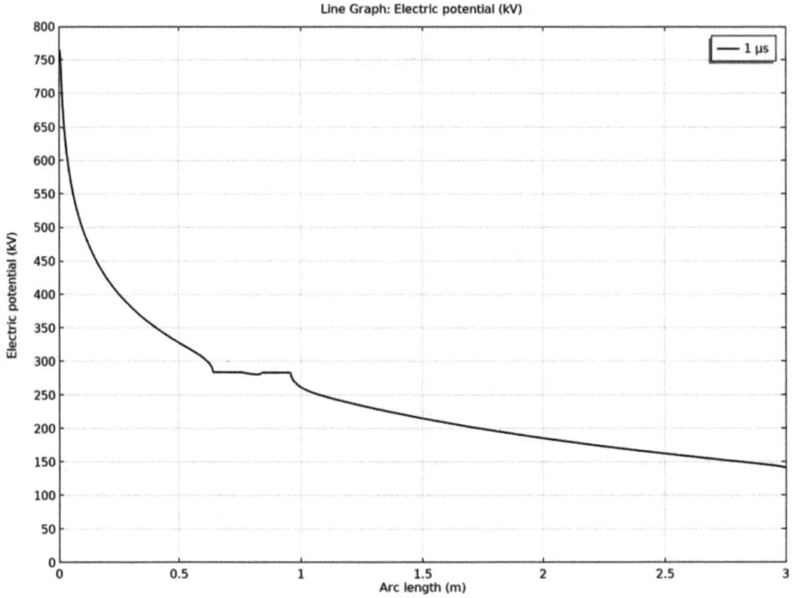

Fig. 7.12 Potential on the earth's surface along line in Fig. 6.9 at the time of the current peak t = 1 μs for 0.25/100

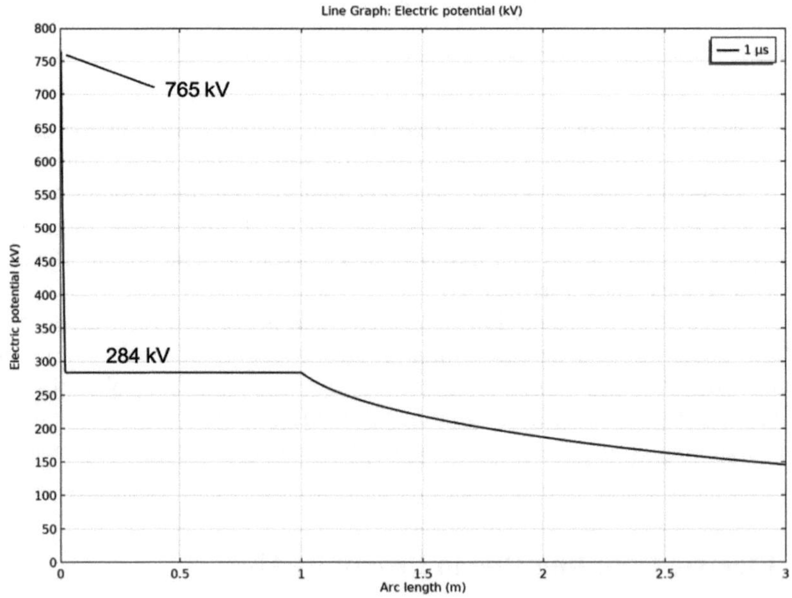

Fig. 7.13 Potential on the earth's surface along line in Fig. 6.13 at the time of the current peak t = 1 μs for 0.25/100

potential difference between the potential at the arrester (position of the hand) and the potential at the hand. This results in ranges similar to those derived in Sect. 6.4.

7.5 Effect of Potential Control with Meshed Earth Grid Underneath of the Asphalt Layer

Due to the limitation of the COMSOL-Multiphysics programme, only the galvanic coupling can be calculated by electrical currents in the time domain. The inductive coupling can be calculated e.g. with the network analysis program Microcap. However, it was also possible to calculate the effect of the inductive coupling with the numerical field calculation with COMSOL-Multiphysics, see Sect. 7.5.2. In order to avoid the hazards, Sect. 7.5.1 examines whether a potential control in the ground below the asphalt layer is expedient to reduce the galvanic coupling. According to Fig. 7.14, the potential control is designed as a mesh grid with a mesh width of 0.5 m and 6 mm stainless steel rods and is located at a frost-proof depth of 0.5 m.

7.5.1 Effect of Potential Control on Galvanic Coupling

7.5.1.1 Person in Contact with Down Conductor

According to Fig. 7.15, only the displacement current flows in the person, the time curve of which is shown in Fig. 7.17 as an example for the lightning current 0.25/100. The galvanic component is negligible and the energy conversion W and charge Q due to galvanic coupling UE (cf. Fig. 2.1) are harmless. The galvanic portion is negligible and the energy conversion W and the charge Q due to galvanic coupling UE (cf. Fig. 2.1) are harmless, Table 7.3. The effect of the mesh grid is clearly shown in the potential curve in Fig. 7.16, the galvanic voltage drop across the person is about 0.16 kV and this results in only a small capacitive displacement current of **6.5 A**. The danger, however, remains the effect of the induced voltage UM, see this chapter. However, the effect of the induced voltage UM remains as a danger, see Sect. 7.5.2. Similar ratios for the energy W and the charge Q also result if the person does not touch the discharge. A detailed recalculation is therefore not necessary.

7.5.1.2 Person not in Contact with the Down Conductor

Because of the small potential difference, there is no flashover from the discharge to the person.

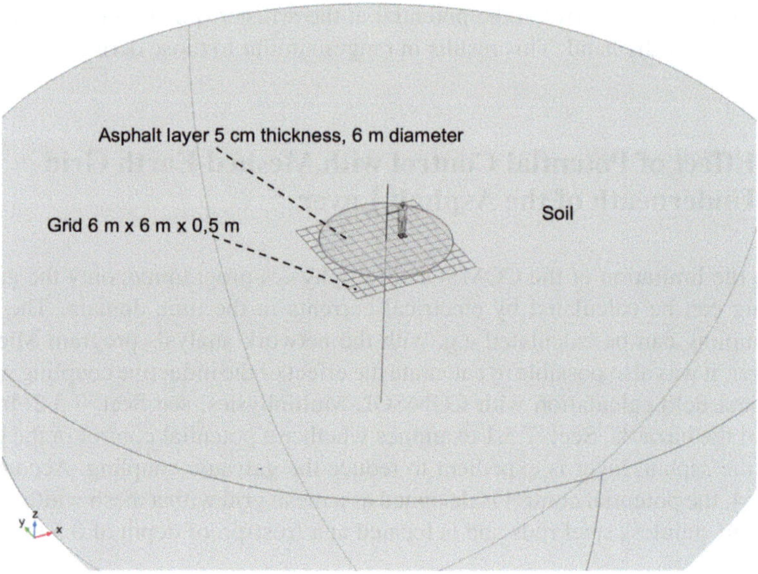

Fig. 7.14 Asphalt layer with a thickness of 5 cm and mesh 6 m × 6 m × 0.5 m with 6 mm stainless steel bars

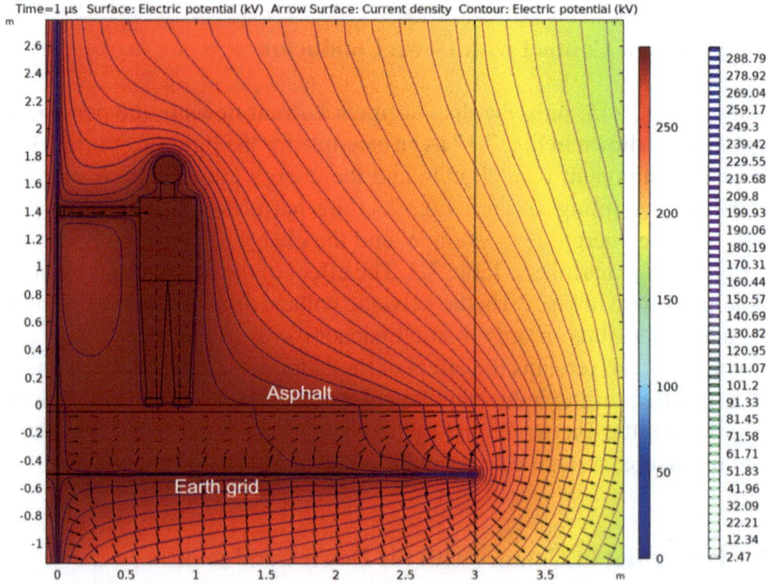

Fig. 7.15 Field at the time of the current peak at t = 1 µs with equipotential surfaces and lines as well as vectors of current density. Negative subsequent stroke 0.25/100 with a peak value of 24.7 kA according to Table 3.1. Soil $\rho_E = 250\ \Omega$m. Current density vectors in logarithmic representation

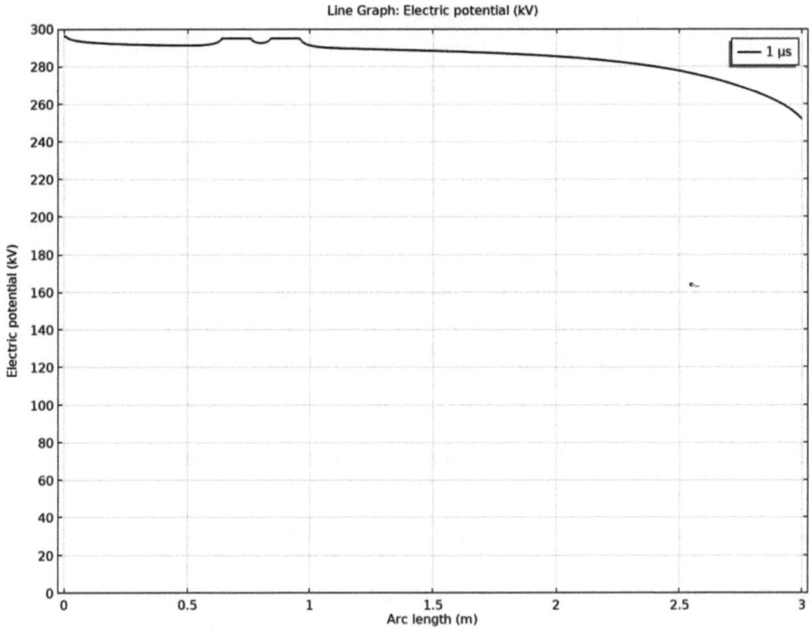

Fig. 7.16 Potential on the earth's surface along line in Fig. 6.9 at the time of the current peak t = 1 µs for 0.25/100

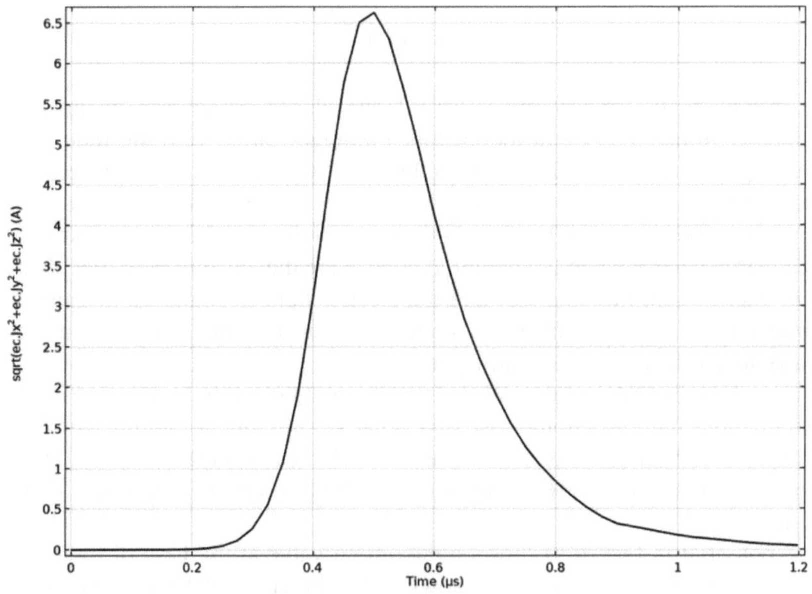

Fig. 7.17 Displacement current (Absolute) through Person for 0.25/100

Table 7.3 Results from the effect of displacement current for a site insulation with wet asphalt with ρ = 10,000 Ωm and potential equalisation. Earth electrode type A; Class I

	I_B	V_{Hand}	V_{Foot}	U_P	I_K	W	Q
	kA	kV	kV	kV	A	Ws	mAs
+10/350	52	666.373	666.365	–	0.5	1e-3	3e-6
−1/200	42	536.98	536.82	–	3.8	7e-3	2.7e-6
−0.25/100	24.7	316.117	315.942	–	6.5	5e-3	1e-6

I_B—Peak value of the injected lightning current according to Table 3.1
I_K—Peak value of the body current at the time of the current peak
V_{Hand}—Potential at the point of contact of the hand against distant earth
V_{Foot}—Potential at the foot against distant earth
U_P—Voltage between hand and foot

7.5.2 *Effect of Potential Control on Inductive Coupling*

7.5.2.1 Person in Contact with the Down Conductor

According to Fig. 2.1, a voltage proportional to the time derivative of the exciting magnetic flux density or the exciting lightning current is induced in the area spanned by the person according to the law of induction. The induced voltage acting in the circuit can be determined by the mutual inductance M as follows.

$$U_{ind} = M * \frac{di}{dt}$$

According to the reciprocity law [32], the closed circuit can be broken and the induced voltage can be introduced as a voltage source with the function U_{ind}. This voltage form is shown in Fig. 19.3. In COMSOL-Multiphysics, the connection between the hand and the rod shown in Fig. 7.18 is separated and the voltage source U_{ind} is inserted between the hand and the down conductor. The down conductor, the vertical earth electrode and the potential control are then at zero potential. Thus, the current distribution, the potential and the electric field strength are investigated in the time domain. The current shows a polarity reversal in Fig. 7.19, which is due to the capacitive effect of the asphalt layer.

The evaluation of the energy W and the charge Q from the results in Fig. 7.18 is shown in Table 7.4. Only in the case of a negative subsequent lightning stroke there is a danger in lightning protection classes I and II due to excessive energy. However, it should be noted that 50% of all negative cloud-to-earth flashes also contain a subsequent stroke.

The potential in Fig. 7.20 shows a maximum value at the base of the person of 98 kV. The field strength in Fig. 7.21 shows peak values up to 80 kV/cm. A comparison with Fig. 20.5 shows, however, that surface discharges at a surge voltage of 1.2/50 at 98 kV are unlikely. Due to the high electric field strength at the base, however, a corona occurs there as a pre discharge. The formation of surface discharges with a

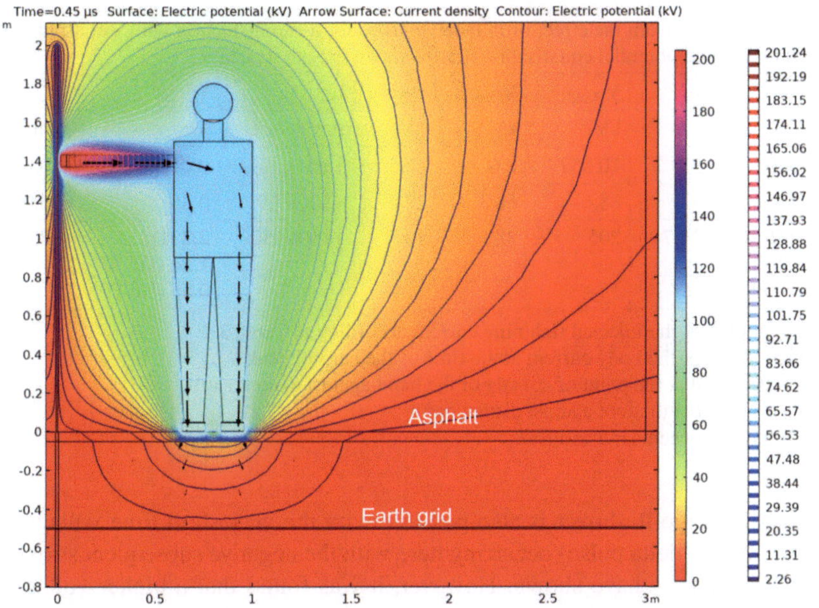

Fig. 7.18 Field at the time of the maximum current slope at t = 0.45 µs with equipotential surface and lines as well as vectors of current density. Negative subsequent stroke 0.25/100 with a peak value of 24.7 kA according to Table 3.1. Soil $\rho_E = 250\ \Omega m$. Asphalt $\rho = 10.000\ \Omega m$, $\varepsilon_r = 3$. Current density vectors in logarithmic representation

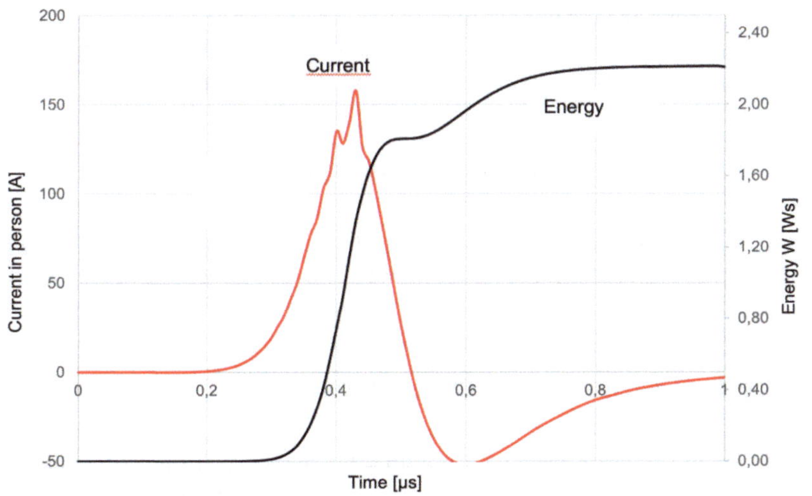

Fig. 7.19 Current (x-component) in the person and energy W converted in the person by the inductive effect of the injected lightning current. Negative sequential flash 0.25/100 with a peak value of 24.7 kA according to Table 3.1. Soil $\rho_E = 250\ \Omega m$. Asphalt $\rho' = 10.000\ \Omega m$, $\varepsilon_r = 3$

Table 7.4 Results from the effect of induced voltage for a site insulation with wet asphalt with ρ = 10,000 Ωm and potential equalisation. Earth electrode type A; Class I

	I_B	V_{Hand}	V_{Foot}	U_P	I_K	W	Q
	kA	kV	kV	kV	A	Ws	mAs
+10/350	52	10.557	8.6	1.95	1.85	0.017	0.0135
−1/200	42	88.6	66.3	22.3	+30/−7.9	0.35	0.015
−0.25/100	24.7	205	98	107	+150/−50	**2.2 (I)** **1.24 (II)** 0.55 (III–IV)	0.018

I_B—Peak value of the injected lightning current according to Table 3.1
I_K—Peak value of the body current at the time of the current peak
V_{Hand}—Potential at the point of contact of the hand against distant earth
V_{Foot}—Potential at the foot against distant earth
U_P—Voltage between hand and foot

surface discharge flashover is also possible with the very short time voltage stress of 100–200 ns (delta pulse) occurring here with the negative subsequent stroke [33], as laboratory tests have shown. However, it was found that a higher voltage than 1.2/50 is required for a delta pulse. The figure of surface discharges with a complete surface discharge flashover at 98 kV cannot therefore be answered unequivocally. This would require a laboratory test. A more detailed explanation of the physics of surface discharges is given in Chap. 20.

Without contact, a flashover can occur between the arrester and the person. Figure 17.7, shows the dangerous zone.

Conclusion: Only with lightning protection classes I and II and negative subsequent stroke is there a danger from excessive energy.

7.5.2.2 Person not in Contact with the Down Conductor

Because of the small potential difference of 98 kV as per Fig. 7.20 there is a small risk of a side flash when a person stays close to the down conductor, see Fig. 17.8.

7.6 Asphalt with a Water Level of 2 cm

According to Fig. 7.22, an asphalt layer with a thickness of 5 cm and a diameter of 6 m is placed on the ground surface. A water layer with a height of 2 cm and a diameter of 10 m is placed on top of it. The asphalt has a resistivity of 10,000 Ωm, which according to Table 5.1 corresponds to asphalt in a wet state. The specific resistance of rainwater is 100 Ωm, Table 5.1.

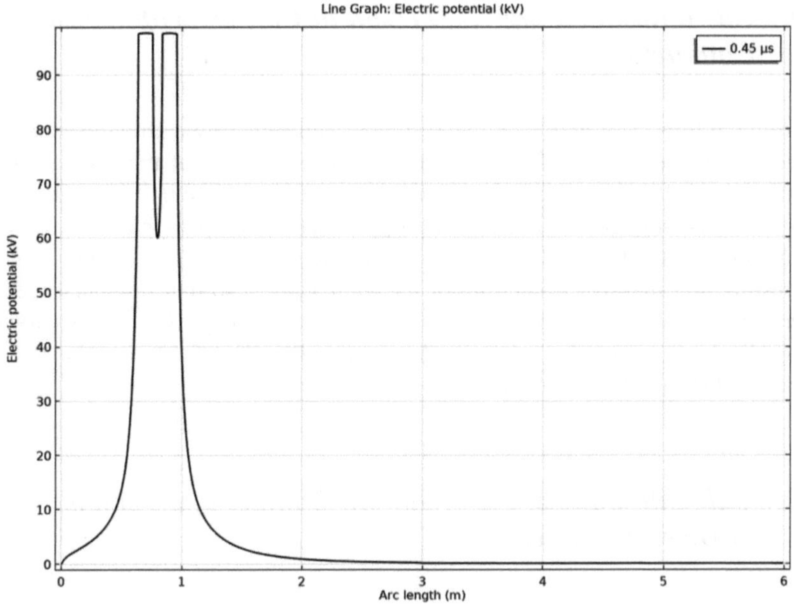

Fig. 7.20 Potential curve on the earth's surface along line in Fig. 6.9 at the time of the current peak t = 0.45 µs for 0.25/100

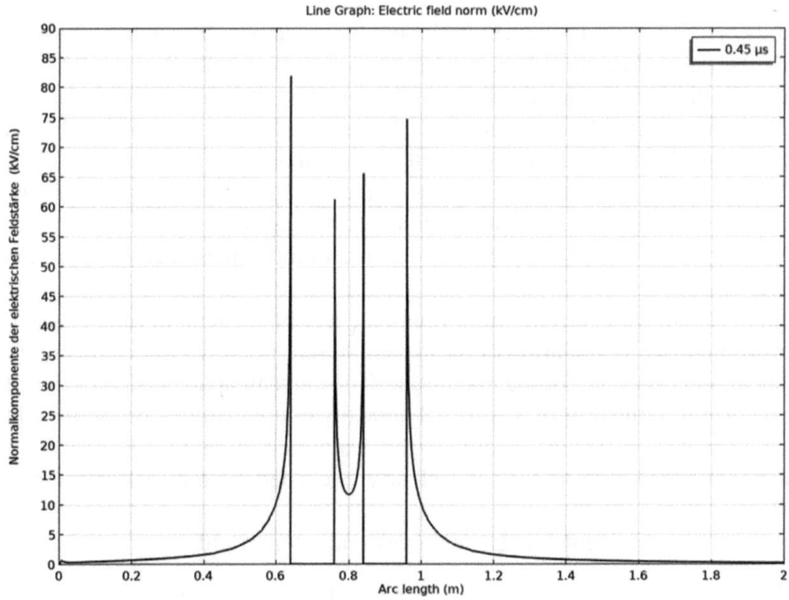

Fig. 7.21 Field strength on the earth's surface along line in Fig. 6.9 at the time of the current peak t = 0.45 µs for 0.25/100. The person has no contact to the down conductor

7.6.1 Galvanic Coupling U_E, Person in Contact with Down Conductor

According to Figs. 7.23 and 7.24, this results in a current flow within the water layer. The course of the earth surface potential along the surface of the asphalt layer is shown in Fig. 7.25. The current density through the centre line in the water is shown in Fig. 7.26. A voltage of 274 kV is dropped across the person and a voltage of 482 kV between the foot and the distant earth. The total effect is shown in Table 7.5. This results in a massive exceeding of the limit values, so that one person dies. It is therefore unnecessary to consider the possibility of gliding discharges. However, gliding discharges can also occur under water [42]. The fields for the lightning currents 1/200 and 10/350 are similar.

7.6.2 Galvanic Coupling U_E Without Contact of a Person to Down Conductor

The calculated arrangement corresponds to Fig. 7.22, whereby the person does not touch the down conductor. Figure 7.27 shows the corresponding field figure. A short current pulse of 3 A flows through the arm, see Fig. 7.31. The potential curve in Fig. 7.28 differs from that in the case with contact of the person in Fig. 7.25. The person therefore has an influence on the potential curve. The potential along a line through the person's arm in Fig. 6.13 is shown in Fig. 7.28. The decisive voltage for dimensioning an insulated down-conductor must therefore be determined from Fig. 7.28 in the case without contact with the person. In this case, the insulated down conductor must be dimensioned for a withstand voltage of at least (762 − 278 kV) = 484 kV. This value is 1.33 times greater than the voltage drop across the person when touched. This results in the following withstand voltages for an isolated discharge in Table 7.6 for the three lightning current forms and the values according to Table 3.1 (Fig. 7.29).

Figure 7.27 shows a current flow through the legs of the person, which is caused by the step voltage. The evaluation of the current height is shown in Fig. 7.30 for the z-component and Fig. 7.31 for the magnitude. The current, apart from the capacitively coupled current in the arm of 3 A, is galvanically coupled and is proportional to the exciting current 0.25/100. According to the evaluation in Table 7.7, the limit values for the permissible step voltage are clearly exceeded here at a lightning current 10/350.

A person who does **not touch** the down conductor would:

(a) Die due to the acting voltage of 484 kV and the associated flashover, for values see Table 7.4.
(b) Die from the effect of the step voltage.

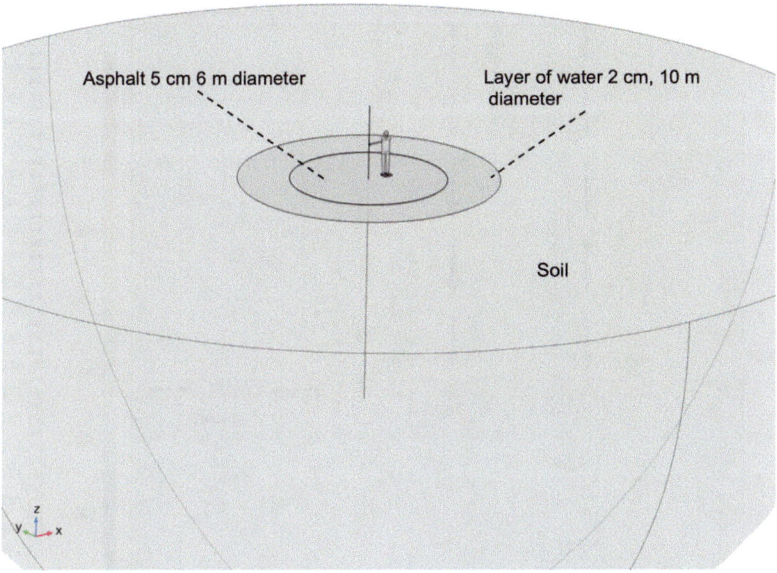

Fig. 7.22 Asphalt layer with a thickness of 5 cm with an overlying water layer of 2 cm thickness

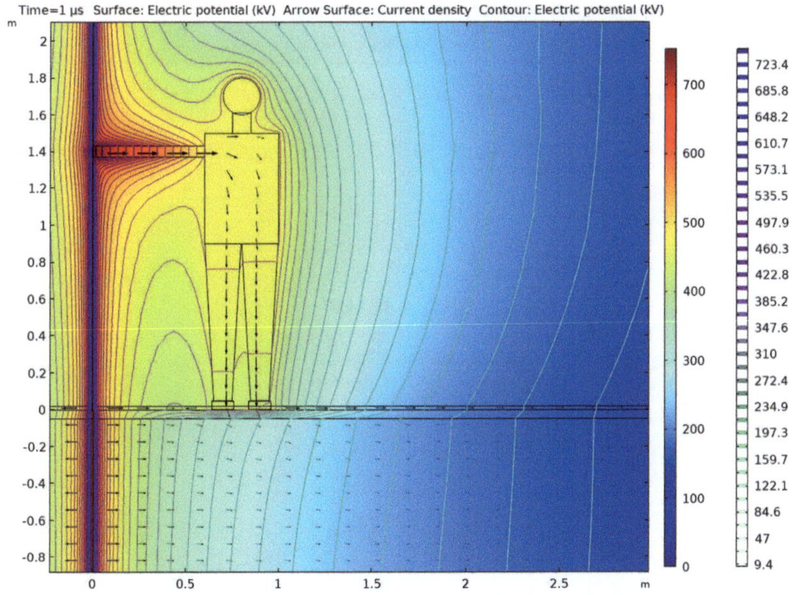

Fig. 7.23 Field at the time of the current peak at t = 1 µs with equipotential surfaces and lines as well as vectors of current density. Negative subsequent stroke 0.25/100 with a peak value of 24.7 kA according to Table 3.1. Soil $\rho_E = 250\ \Omega m$. Water $\rho_W = 100\ \Omega m$, $\varepsilon_r = 80$. Asphalt $\rho = 10.000\ \Omega m$, $\varepsilon_r = 3$. Current density vectors in logarithmic representation

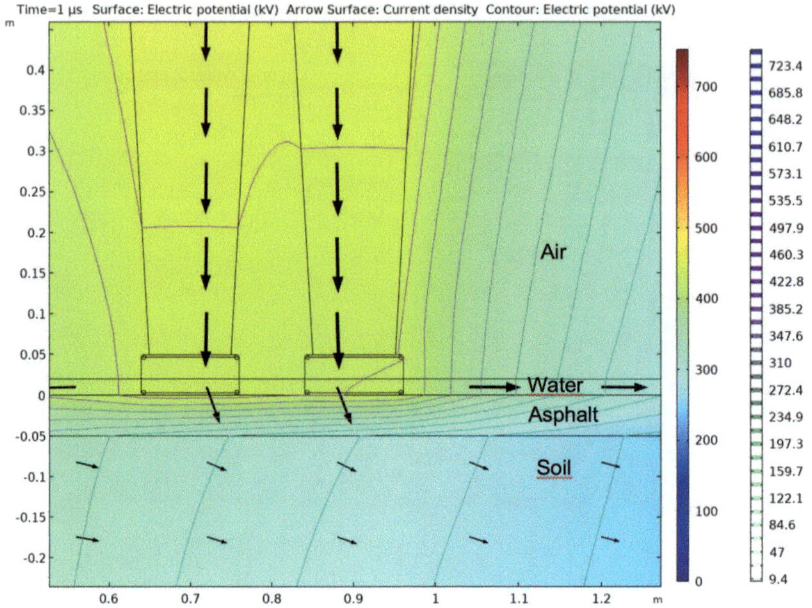

Fig. 7.24 Extended view of Fig. 7.23

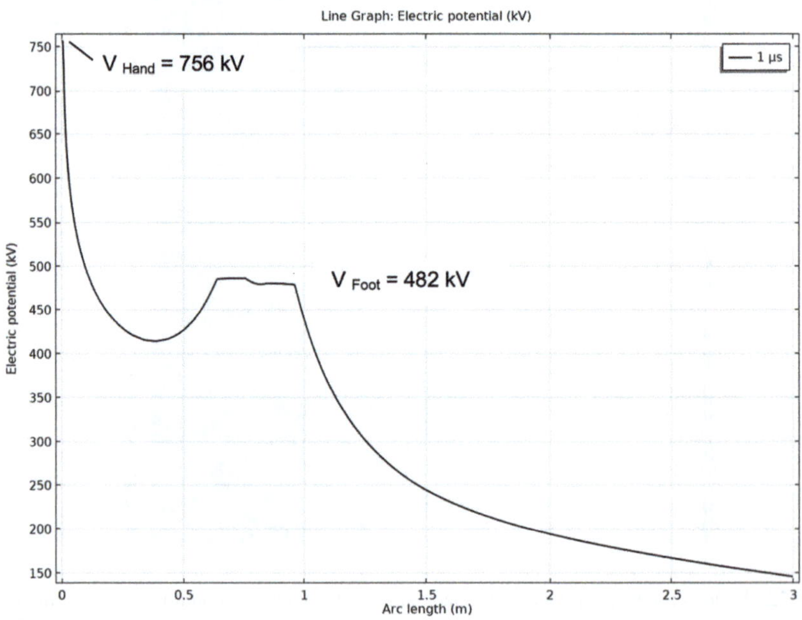

Fig. 7.25 Potential on the earth's surface along line in Fig. 6.9 at the time of the current peak t = 1 μs for 0.25/100

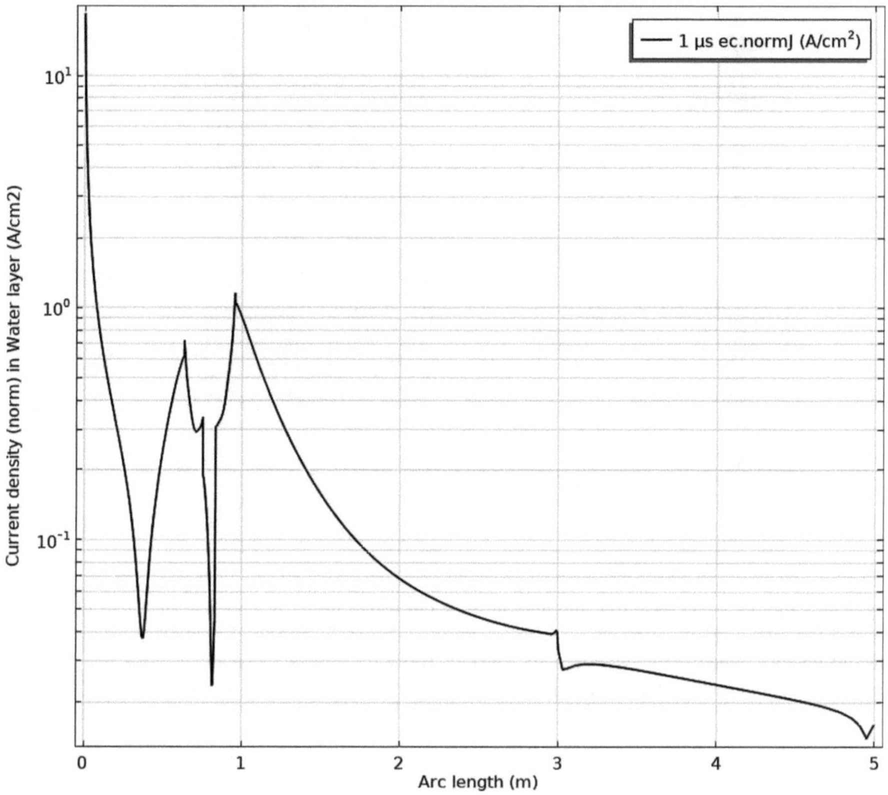

Fig. 7.26 Current density along the centre line through the water layer at the time of the current peak t = 1 μs for 0.25/100

Table 7.5 Results with a layer of water on asphalt. Earth electrode type A; Class I. Exceeded limit values are shown in bold. The limit values are also exceeded in Class II and III–IV

	I_B	V_{Hand}	V_{Foot}	U_P	I_K	W	Q
	kA	kV	kV	kV	A	Ws	mAs
+10/350	52	1591	1020	571	571	**83.640**	**282**
−1/200	42	1287.5	828	459.5	463	**30.982**	**133**
−0.25/100	24.7	756	482	274	274	**5.412**	**40**

I_B—Peak value of the injected lightning current according to Table 3.1
I_K—Peak value of the body current at the time of the current peak
V_{Hand}—Potential at the point of contact of the hand against distant earth
V_{Foot}—Potential at the foot against distant earth
U_P—Voltage between hand and foot

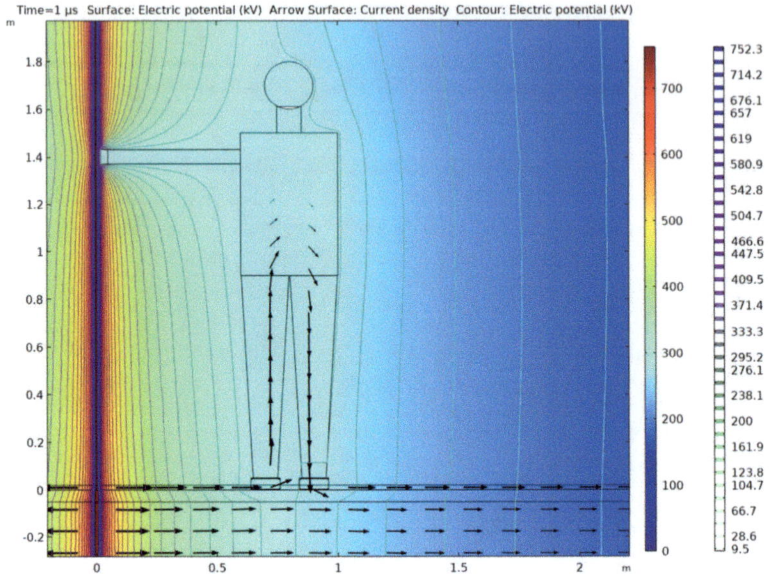

Fig. 7.27 Field for the case that the person does not touch the down conductor. Field plot at the time of the current peak at t = 1 µs with equipotential surface and lines as well as vectors of the current density. Negative subsequent stroke 0.25/100 with a peak value of 24.7 kA according to Table 3.1. Soil $\rho_E = 250\ \Omega$m. Current density vectors in logarithmic representation

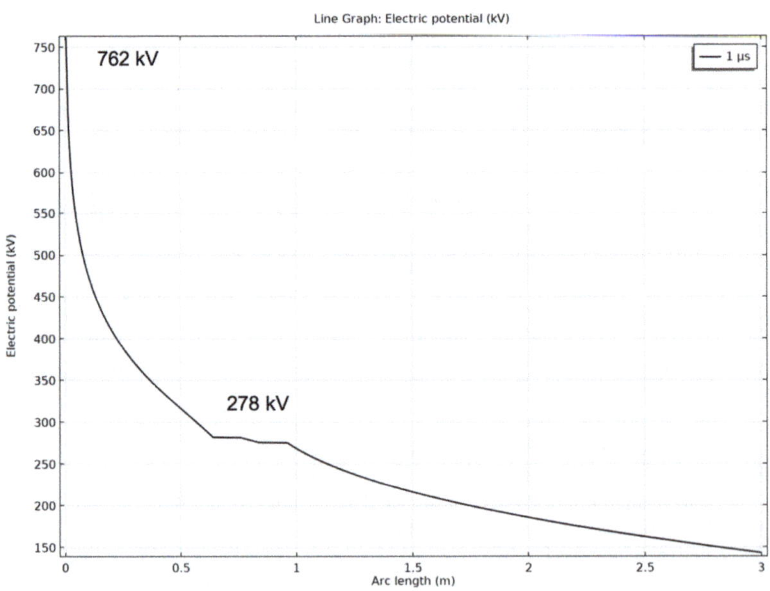

Fig. 7.28 Potential curve on the earth's surface along line in Fig. 6.9 at the time of the current peak t = 1 µs for 0.25/100

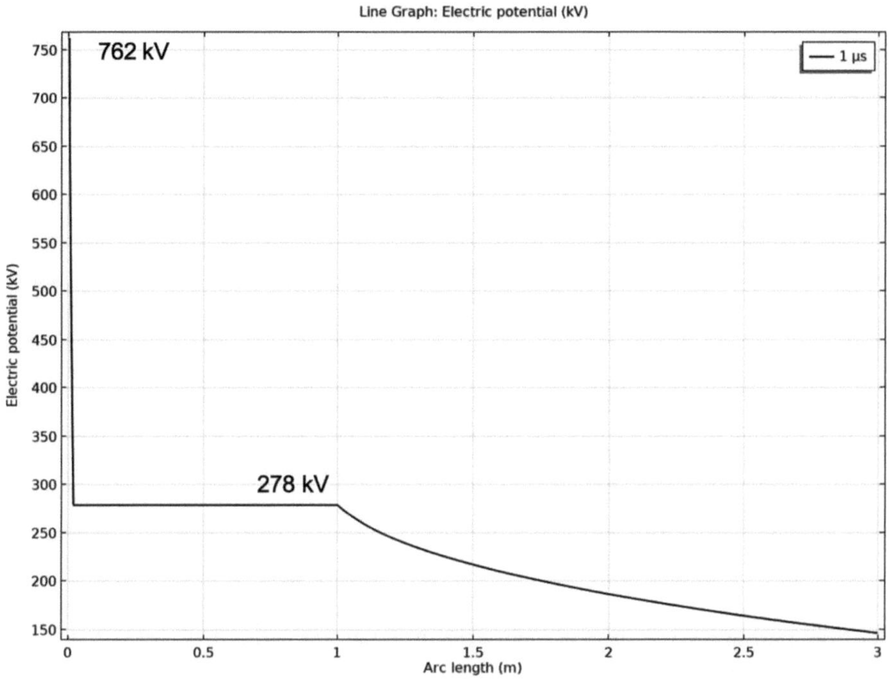

Fig. 7.29 Potential curve on the earth's surface along line in Fig. 6.13 at the time of the current peak t = 1 µs for 0.25/100

Table 7.6 Results for the case without touching the down conductor. Type A earth electrode; Class I

	U_P	U_S	U_S/U_P
	kV	kV	
10/350	767	1000	1.30
1/200	621	814	1.31
0.25/100	362	484	1.33

7.6.3 *Inductive Coupling U_M*

Due to the massive exceeding of the limit values, see Table 7.6, a calculation of the inductive coupling is unnecessary here. However, the inductive coupling is important again for the case of potential control and is therefore only examined for this case in Sect. 7.7.2.

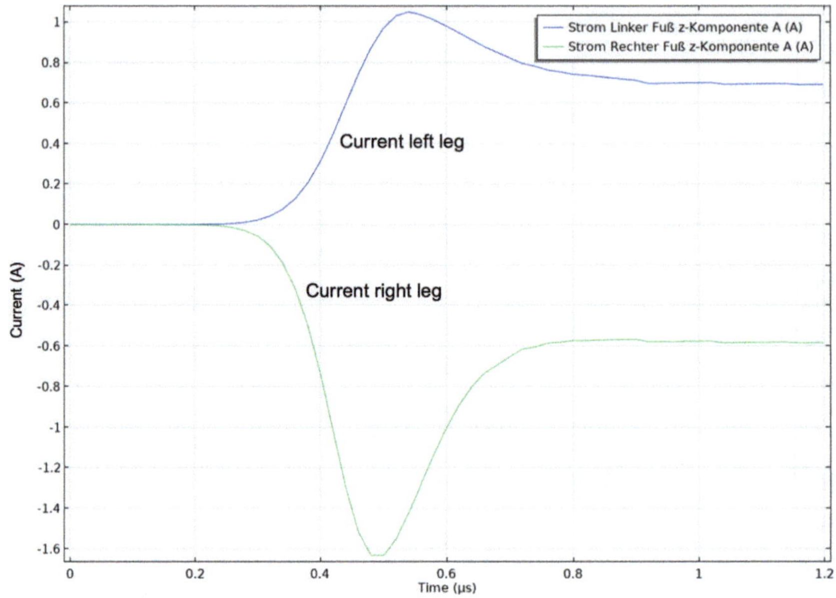

Fig. 7.30 Current (z-component) in both legs for 0.25/100. The person does not touch the down conductor. Y-scale in A

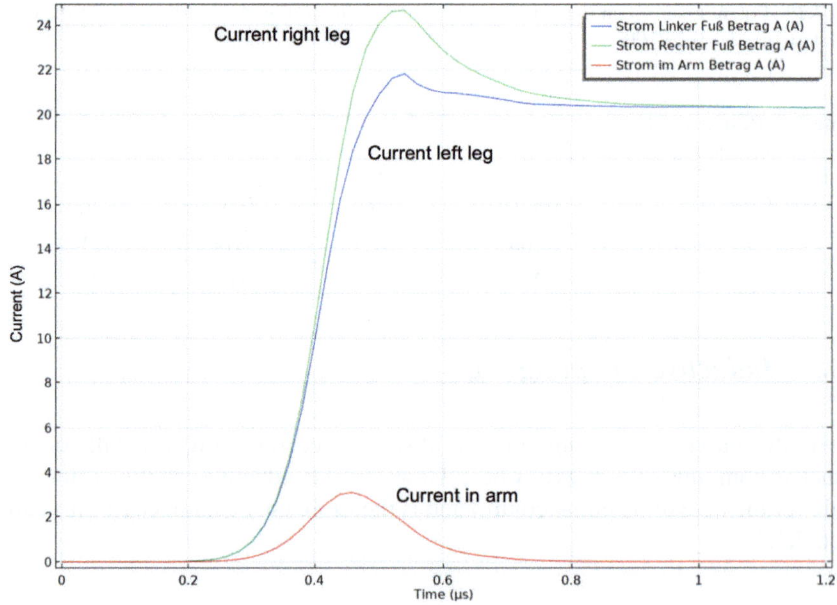

Fig. 7.31 Current (magnitude) in both legs for 0.25/100. The person does not touch the down conductor. Y-scale in A

Table 7.7 Current level I. Earth electrode type A; Class I

	I_B	I	Limit value for step voltage according to [34]
	kA	A	A
10/350	52	46.3	25
1/200	42	37.4	No fixed value available
0.25/100	24.7	22	

I_B—Peak value of the injected lightning current according Table 3.1
I—Peak value of the current in the body of a person

7.7 Effect of Potential Control Measures Under a Water Covered Asphalt Layer

A potential equalisation with a grid of 6 m × 6 m with a mesh size of 0.5 m and 6 mm stainless steel rods is located at a frost proof depth of 0.5 m below the person according to Fig. 7.32. In the following, the influence of the grid on the galvanic and inductive coupling is calculated with the programme COMSOL-Multiphysics.

7.7.1 Effect of Potential Control Measures on Galvanic Coupling

7.7.1.1 Person is in Contact with Down Conductor

The field in Fig. 7.33 shows a lower current in the person than in the case without potential control. The potential curve in Fig. 7.34 along the earth's surface is much flatter compared to Fig. 7.35 and the maximum value is, as expected, significantly reduced. The vectors of the current density shown in the person are essentially given by the displacement current of 6.9 A peak value, compare Fig. 7.17. A voltage with a peak value of 3.8 kV drops across the person due to the galvanic coupling and a current of 3.8 A of the form 0.25/100 flows. The resulting energy W and the charge Q are shown in Table 7.8 and exceed the limit values despite the potential control. In addition, there is the effect of the induced voltage, which is examined in Sect. 7.7.2. Similar conditions also result if the person does not touch the discharge. A detailed recalculation is therefore not necessary.

7.7.1.2 Person is not in Contact with Down Conductor

Because of the small potential difference, there is no flashover from the down conductor to the person.

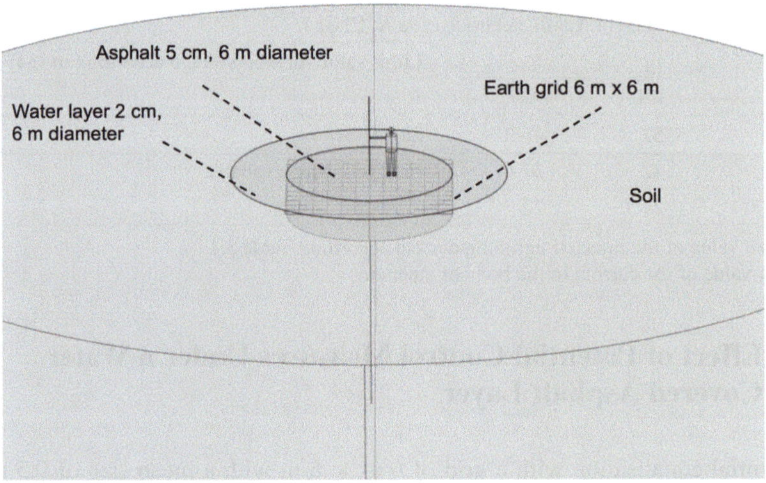

Fig. 7.32 Asphalt layer with a thickness of 5 cm with a water layer of 2 cm thickness on top and additional potential control

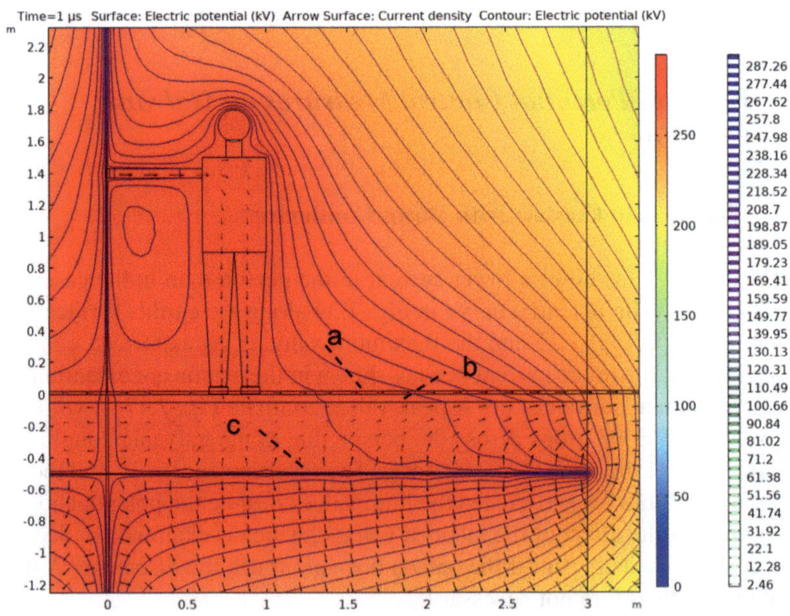

Fig. 7.33 Field at the time of the current peak at t = 1 μs with equipotential surfaces and lines as well as vectors of current density. Negative subsequent stroke 0.25/100 with a peak value of 24.7 kA according to Table 3.1. Soil $\rho_E = 250\ \Omega m$. Water $\rho_W = 100\ \Omega m$, $\varepsilon_r = 80$. Asphalt $\rho = 10.000\ \Omega m$, $\varepsilon_r = 3$. Current density vectors in logarithmic representation. **a** Layer of water 2 cm; **b** Asphalt 5 cm; **c** Grid 0.5 m × 0.5 m × 6 mm

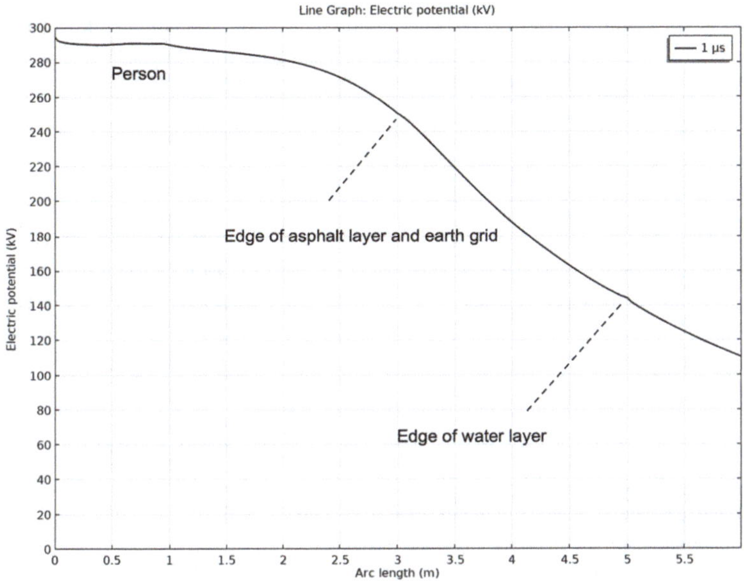

Fig. 7.34 Potential on the earth's surface along line in Fig. 6.9 at the time of the current peak t = 1 µs for 0.25/100

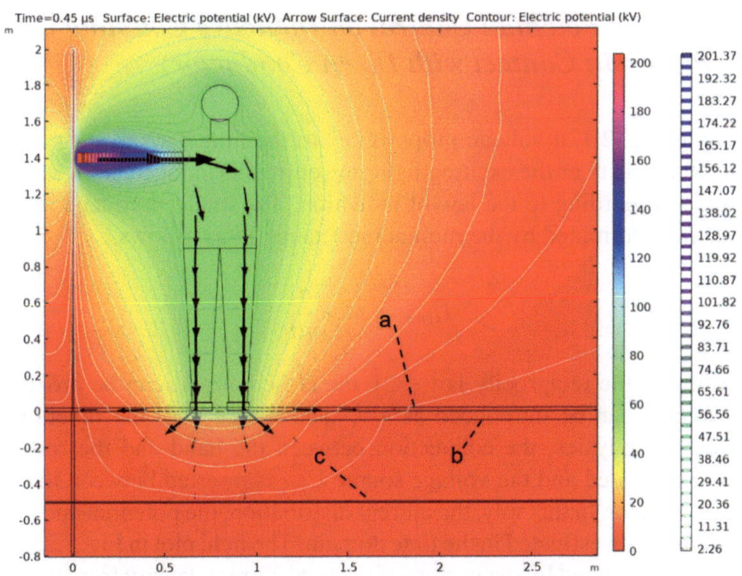

Fig. 7.35 Field plot at time of maximum current slope at t = 0.45 µs with equipotential surface and lines as well as vectors of current density. Negative subsequent stroke 0.25/100 with a peak value of 24.7 kA according to Table 3.1. Soil $\rho_E = 250\ \Omega$m. Current density vectors in logarithmic representation. **a**, **b**, **c**: see Fig. 7.34

Table 7.8 Results from the effect of galvanic coupling for a site insulation with wet asphalt with $\rho = 10{,}000\ \Omega m$ and equipotential bonding with a 2 cm high water layer. Type A earth electrode; Class I. Values for other lightning protection classes in brackets Values that are exceeded are printed in bold

	I_B	V_{Hand}	V_{Foot}	U_P	I_K	I_{KG}	W from I_{KG}	Q from I_{KG}
	kA	kV	kV		A	A	Ws	mAs
+10/350	52	666.373	666.365	–	0.5	8.05	**16.6 (I)** **9.33 (II)** **4.15 (III–IV)**	**4 (I)** **3 (II)** **2 (III–IV)**
−1/200	42	536.98	536.82	–	3.8	6.5	**6.1 (I)** **3.43 (II)** **1.52 (III–IV)**	**1.9 (I)** **1.06 (II)** 0.475 (III–IV)
−0.25/100	24.7	316.117	315.942	–	6.5	3.8	**1.04 (I)** 0.59 (II) 0.26 (III–IV)	0.5

I_B—Peak value of the injected lightning current according to Table 3.1
I_K—Peak value of the body current at the time of the current peak
I_{KG}—Peak value of the galvanically injected current
V_{Hand}—Potential at the point of contact of the hand against distant earth
V_{Foot}—Potential at the foot against distant earth
U_P—Voltage between hand and foot

7.7.2 *Effect of Potential Control on Inductive Coupling (Person in Contact with Down Conductor)*

According to Fig. 2.1, a voltage proportional to the time derivative of the exciting magnetic flux density or the exciting lightning current is induced in the area spanned by the person according to the law of induction. The induced voltage acting in the circuit can be determined by the mutual inductance M as follows.

$$U_{ind} = M * \frac{di}{dt}$$

According to the reciprocity law [32], the closed circuit can be broken and the induced voltage can be introduced as a voltage source with the function U_{ind}. In COMSOL-Multiphysics, the connection between the hand and the rod shown in Fig. 7.32 is separated and the voltage source U_{ind} is inserted between the hand and the down conductor. In this way, the current distribution, the potential and the electric field strength are investigated in the time domain. The field plot in Fig. 7.35 shows the current flow represented by the vectors of the total current density from displacement current and conduction current. Because of the very high voltage slope, a current also flows through the asphalt layer. The potential flow through the centre of the 2 cm high water surface in Fig. 7.36 shows a potential of about 52 kV at the feet of the person. There are no large gradients, so the electric field strength remains uncritical and therefore no surface discharges are possible. The voltage drop between the feet and

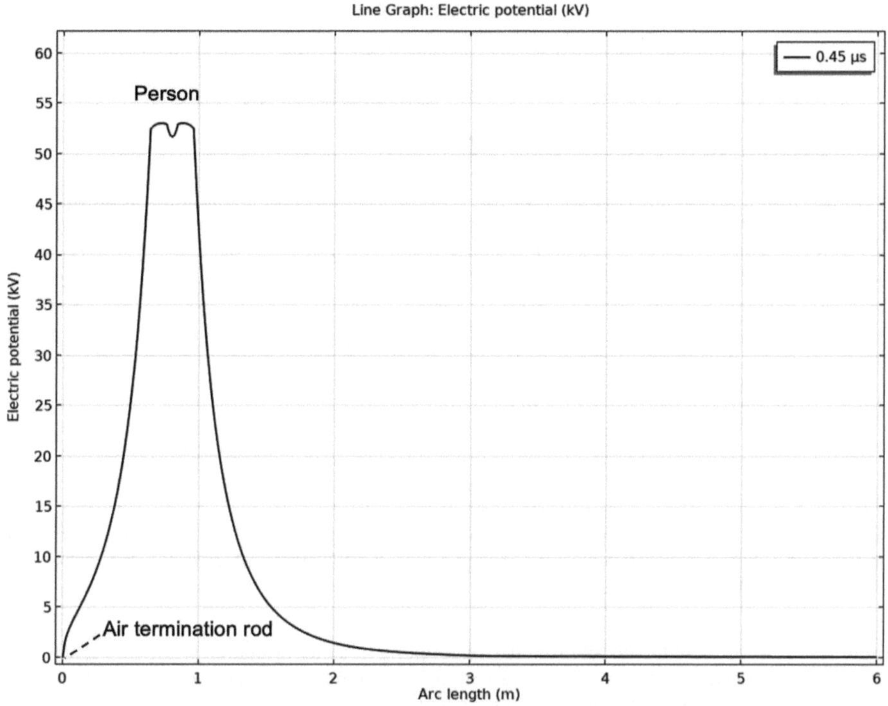

Fig. 7.36 Potential curve through the centre of the water surface at the time of the maximum current slope t = 0.45 μs for 0.25/100

the down conductor reduces the current in the person compared to the case when the person would stand on an ideal equipotential surface. The current flow through the person is shown in Fig. 7.37 and reaches a peak value of 176 A, whereby the average pulse width is only 0.15 μs. This results in the energy converted in the person, also shown in Fig. 7.37, with a final value of 3.18 Ws, which exceeds the permissible limit of 1 Ws. The charge is below the limit of 1 mAs. Table 7.9 shows the overview of the results also for the lightning currents 1/200 and 10/350. In lightning protection class I, both the negative first stroke and the negative subsequent stroke are dangerous. The 10/350 lightning current, on the other hand, is not dangerous. The value for the energy W can be converted quadratically to the other lightning protection classes. However, it should be noted that 50% of all negative cloud-to-ground flashes also contain a negative subsequent stroke. Then, according to the current state of science, the energy components have a cumulative effect, so that there is actually a danger in all classes for combined stress. On the other hand, there is a greater danger due to galvanic coupling in Sect. 7.7.1. For physical reasons, both effects, galvanic coupling and inductive coupling, act simultaneously, so that in all lightning current forms and lightning protection Classes, the permissible limit values are exceeded and other measures must be used to solve the problem, which are described in more detail in Chap. 12.

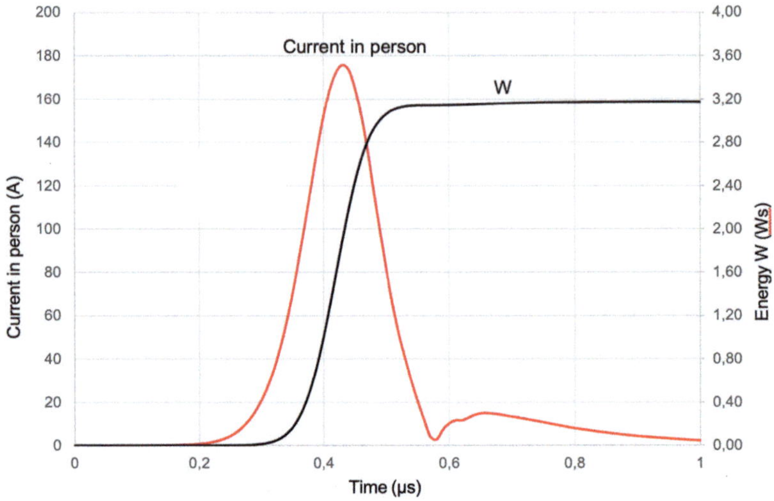

Fig. 7.37 Current in the person and energy W converted in the person by the inductive effect of the injected lightning current. Negative subsequent stroke 0.25/100 with a peak value of 24.7 kA according to Table 3.1. Soil $\rho_E = 250\ \Omega m$

Table 7.9 Results from the effect of inductive coupling in a site insulation with wet asphalt with $\rho = 10{,}000\ \Omega m$ and equipotential bonding with a 2 cm high water layer. Earth electrode type A; Class I. Values for other lightning protection classes in brackets

	I_B	U_{ind}	V_{Hand}	V_{Foot}	U_P	I_{KM}	W from I_{KM}	Q from I_{KM}
	kA	kV	kV	kV	kV	A	Ws	mAs
+10/350	52	10.85	10.85	4.6	6.25	6.22	0.2	0.0475
−1/200	42	86.9	86.9[a]	29.4	57.5	58.6	**1.49 (I)** 0.83 (II) 0.37 (III–IV)	0.037
−0.25/100	24.7	203	203[a]	52.6	150.4	176	**3.18 (I)** **1.78 (II)** 0.8 (III–IV)	0.029

I_B—Peak value of the injected lightning current according to Table 3.1
I_{KM}—Peak value of the body current due to magnetic coupling at the time of the current peak
V_{Hand}—Potential at the point of contact of the hand against distant earth
V_{Foot}—Potential at the foot against distant earth
U_P—Voltage between hand and foot
U_{ind}—Induced voltage
[a] The difference between potential at the hand and at the foot is also influenced by the capacitive displacement component. Therefore, the voltage drop at the person is not determined by Ohm's law alone

7.8 Evaluation of the Results of Site Insulation with Asphalt

An asphalt layer reduces the load on a person. However, the load values in Table 7.1 are still too high. With a wet asphalt layer, the danger is determined by the galvanic coupling. With a completely dry and highly insulating asphalt layer, the danger is caused by the formation of surface discharges. In the case of a surface discharge flashover, the distance between the base and the discharge is bridged and then the conditions as in Table 6.7 apply.

A potential control reduces the load below the limit values according to Tables 7.3 and 7.4. The voltage values at the person (hand and foot) are significantly reduced. The electric field strength at the foot of a person exceeds the limit value of the electric strength of the air and surface discharges occur. In this case, too, the insulation effect of the asphalt and the effect of the potential control become ineffective in the event of a surface discharge flashover, and conditions according to Table 6.7 apply that are lethal.

In the case of a 2 cm high water layer on the asphalt, the limit values are massively exceeded, Table 7.5. Potential control is a necessary, but not sufficient solution in all lightning protection classes, as the limit values are still exceeded, Tables 7.8 and 7.9. Furthermore, it must be taken into account in risk assessments that negative cloud-to-earth lightning occurs in approx. 50% of all cases. The values converted in the person must be assumed to be cumulative.

Overall, site isolation is therefore a necessary but not sufficient measure in all cases. Potential control is another necessary and partly sufficient measure. A comprehensive and sufficient solution for all lightning protection classes is an isolated down conductor, which is described in Chap. 12.

Chapter 8
Effect of Site Insulation with Gravel According to IEC 62305-3

The effect of site insulation with gravel is probably similar to the effect of a water permeable layer. A detailed calculation is therefore not carried out. One solution, even in the case of gravel and potential equalisation, is to install an insulated down conductor. Investigations by Beierl[1] have also shown that gravel is not suitable as an insulating material because of the risk of flashovers.

[1] Beierl O.: Isolationsfestigkeit bei Bodenuntergründen (Forschungsprojekt 76). Vortrag während der Anwenderfachtagung des VDE/ABB am 23.6.2017 in Frankfurt.

J. Meppelink, *The Lightning Rod as a Danger*,
https://doi.org/10.1007/978-3-031-30434-7_8

Chapter 9
Water-Permeable Site Insulation

9.1 Water-Permeable Materials

Data from relevant building materials in [2] are the basis for the following calculations. Water-permeable mineral mixtures are bonded with a polyurethane binder in such a way that a stable, water- and air-permeable top layer is created. Figure 9.2 shows the structure of such a layer, which can only be numerically reproduced globally by an estimated equivalent value of the specific resistance. The parameters required from a high-voltage point of view, such as specific surface capacitance and breakdown voltage, are given in the individual calculated cases (Table 9.1 and Fig. 9.1).

The breakdown in gases, in liquids and in solid bodies is described in the literature in the individual cases as far as possible. For a rod-plate-air spark gap in air, data are given in [21]; for sand in [22] and compiled in Table 9.2.

Soil is a combination of all three states of matter, and a uniform theory for breakdown in soil and its dielectric strength is not known. The composition of soil is shown in Fig. 9.1. Gases can also be dissolved in water. How well individual sand or rock layers conduct electricity depends on their water content, their chemical composition and, above all, their degree of mineralisation: (microbial decomposition of organic substances to inorganic compounds with the release of carbon dioxide). The resistivity of sand depends very much on its moisture content, [20] see Table 9.3.

Mineralised rocks containing ores, graphite and many sulphides are very good underground conductors. In addition to the electrical conduction mechanism in solid bodies, the conductivity of liquids containing salts is therefore also important (e.g. salt water is a good electrical conductor). If, on the other hand, the current density increases to such an extent that the surrounding gases are ionised, this leads to electrical breakdown of these gas components and thus to the ionisation of the soil. In this process, current paths with low impedance practically form parallel to areas with conventional conductivity of solid bodies. Investigations by Berger [22] show:

In dry soil (sand) the breakdown occurs in the pores, i.e. in air as pure gas breakdown according to the streamer mechanism. Evidence of this theory can be

J. Meppelink, *The Lightning Rod as a Danger*,
https://doi.org/10.1007/978-3-031-30434-7_9

Table 9.1 Breakdown field strength of a rod-plate arrangement in air, compared with sand

Shape of voltage	1.2/50		150/2500	
Polarity	+	–	+	–
	kV/cm			
Rod-plane E 50%. 20 cm in air [21]	6.5	14.3	5.6	20
Rod-plane in Sand mit grain size < 0.3 mm [22]. E 0%	5.7	No information		

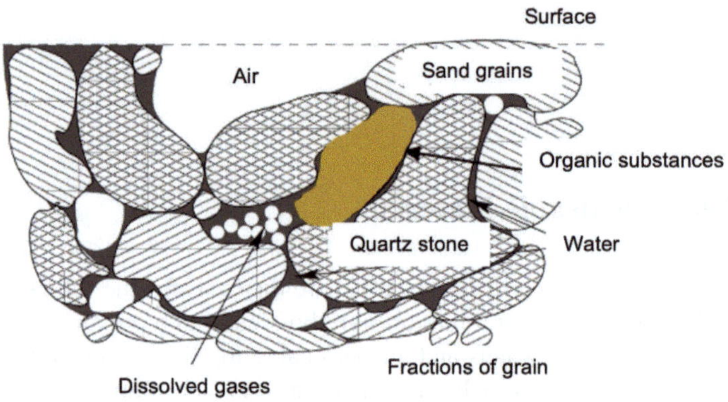

Fig. 9.1 Model of a soil sample. *Source* [23]

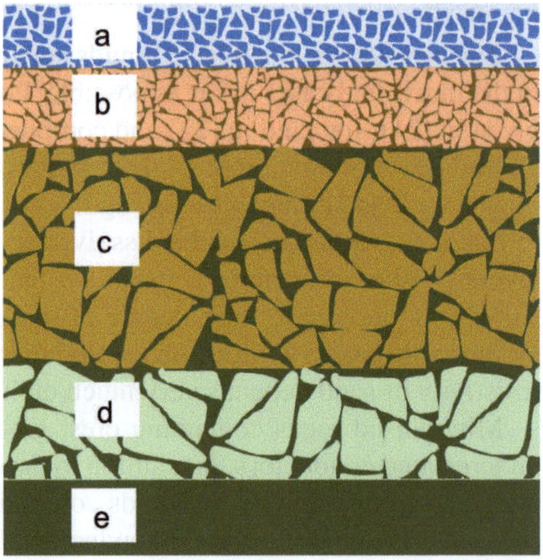

Fig. 9.2 Exemplary structure. Layer **a** 4–5 cm water- and air-permeable mixture of stone and plastics, **b** 6 cm mineral mixture, **c** 15–20 cm base layer of crushed stone, **d** 10 cm clean layer, **e** soil

found by drying samples of soil, evacuating them and filling them with SF_6-gas (Sulphur Hexafluoride) instead of air. It could be shown that the strength increased significantly.

Table 9.2 Specific resistance of sand with different moisture contents

	Humidity	Specific resistance
	%	Ωm
Sand, grain size 0.3 mm	0	31.291
	3	645
	6	530
Sand, grain size 10–12 mm	0	48.466
	2	4.478
	4	1.295

Table 9.3 Parameter of a mixture

		Water	Stone
ρ	Ωm	45	$2 * 10^6$
ε_r		80	8

In the case of wet material (borderline case of pure water), the breakdown only takes place after the water has heated up, so that the breakdown also takes place in air bubbles there. The breakdown field strength decreases with increasing time of halve value of the current pulse. In fact, it is also described that gas bubbles appear before the breakdown during water breakdown.

The average breakdown field strength of a sphere—sphere arrangement is reduced to about 1/3 compared to an equal air gap.

For the evaluation of calculated field strengths, an estimate must be made in each individual case according to the data given above. For future consideration in the normative range, further metrological investigations are recommended.

The effective permittivity ε_{eff} of a mixture of stone and water components required for the simulation is determined according to [26]. The effective resistivity ρ_{eff} results elementary, Fig. 9.3.

9.2 Effect of a Water-Permeable Layer

In the following, a water-permeable layer with a total thickness of 40 cm with the material data according to Fig. 9.3 for a mixing ratio of 10% water and 90% stone for a single discharge of earth electrode type A in Fig. 9.4 is examined with the numerical field calculation as an example according to Fig. 9.2.

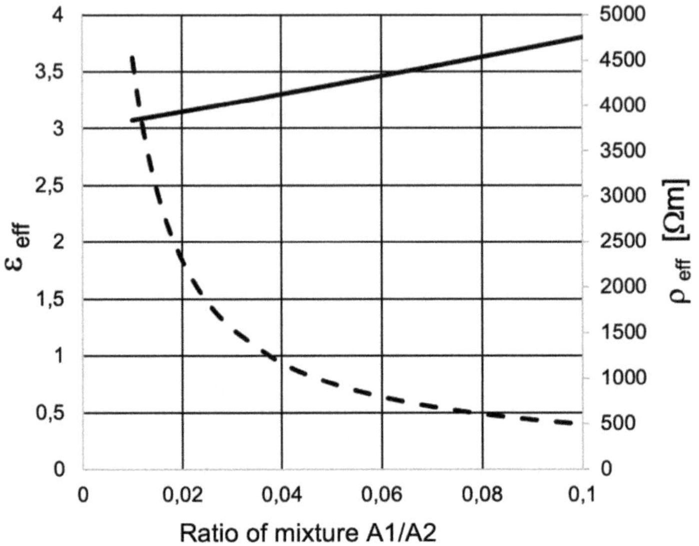

Fig. 9.3 Effective permittivity ε_{eff} (dotted line) and effective resistivity ρ_{eff} (solid line) of a mixture of stone with proportion A2 and water with proportion A1

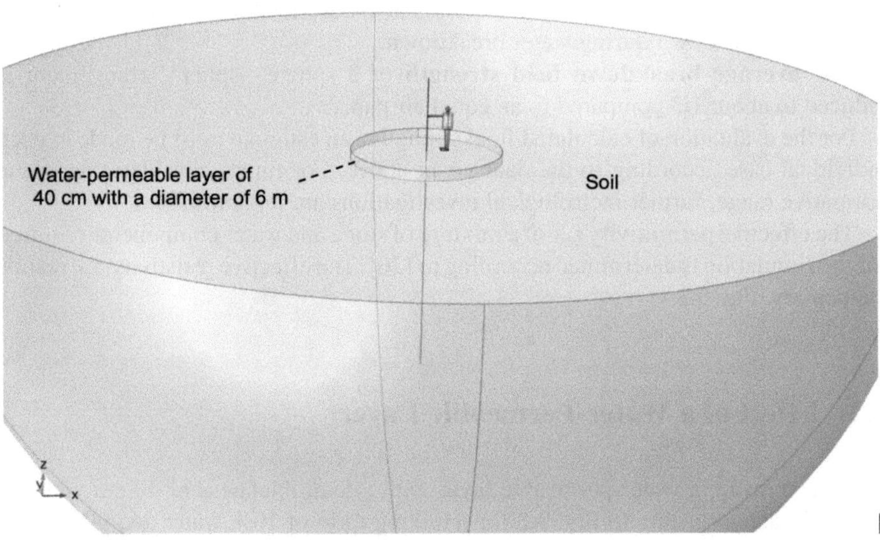

Fig. 9.4 Water-permeable layer with a thickness of 40 cm

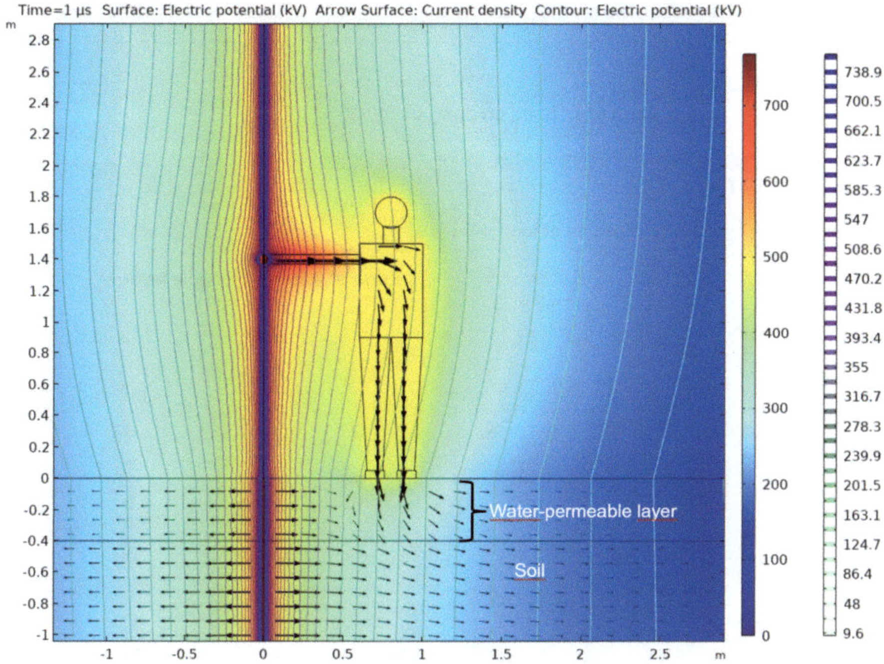

Fig. 9.5 Field of the array with equipotential lines and current density vectors for the load values according to Table 3.1. Negative subsequent stroke 24.7 kA, at the time of the current peak. Water-permeable layer, 40 cm depth, 3 m width, $\rho = 450\ \Omega\text{m}$, $\varepsilon_r = 3.75$, soil: $\rho_E = 250\ \Omega\text{m}$. Current density vectors in logarithmic representation

9.2.1 Galvanic Coupling U_E with a Person in Contact with Down Conductor

The field plot in Fig. 9.5 shows the equipotential surfaces, equipotential lines and current density vectors of the electric current field in the body and in the adjacent layers. Figure 9.6 shows the potential curve on the surface of the water-permeable layer along the line shown in Fig. 6.9 for the lightning current 0.25/100. A person touching a down conductor would almost certainly die in all lightning protection classes. An evaluation of the energy and charge is shown in Table 9.4.

9.2.2 Galvanic Coupling U_E Without a Person in Contact with Down Conductor

The calculated arrangement corresponds to Fig. 9.4, where the person does not touch the down conductor. Figure 9.7 shows the corresponding field plot. A short current

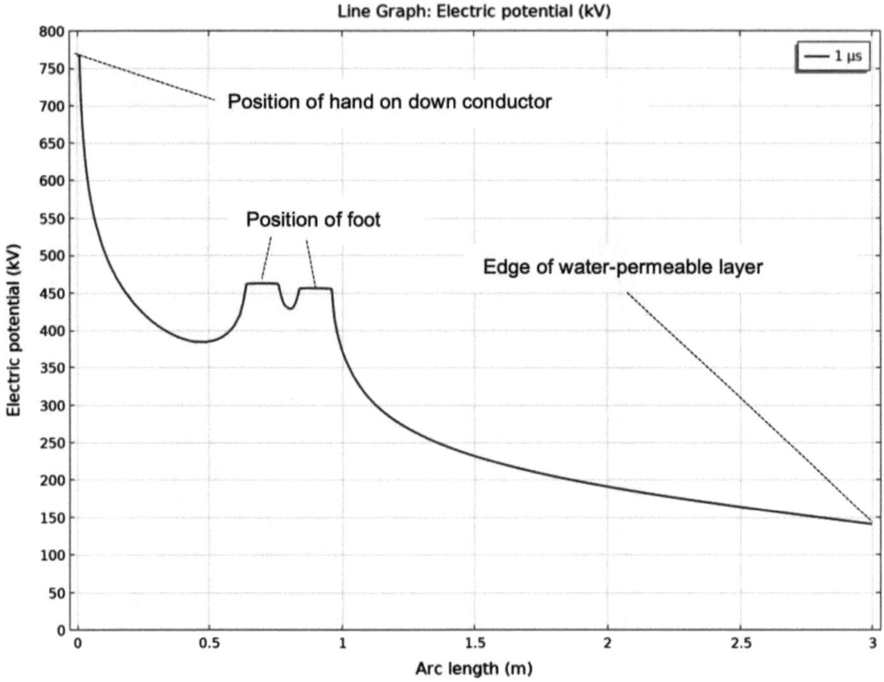

Fig. 9.6 Potential curve along line in Fig. 6.9 at the time of the current peak in the body for lightning current 0.25/100; 24.7 kA peak value

Table 9.4 W, Q with water-permeable layer. Earth electrode type A; Class I. Person touched at a height of L = 1.5 m at a distance of 0.8 m, exceeded limit values are shown in bold

	I_B	V_{Hand}	V_{Foot}	U_P	I_K	W	Q
	kA	kV	kV	kV	A	Ws	mAs
+10/350	52	1580	930	650	650	**108.386**	**321**
−1/200	42	1263	750	513	513	**38.035**	**147**
−0.25/100	24.7	750	440	310	310	**6.927**	**44**

I_B—Peak value of the injected lightning current according to Table 3.1
I_{KM}—Peak value of the body current due to magnetic coupling at the time of the current peak
V_{Hand}—Potential at the point of contact of the hand against distant earth
V_{Foot}—Potential at the foot against distant earth
U_P—Voltage between hand and foot

pulse of 3 A flows through the arm. The potential in Fig. 9.8 differs from that in the case with the person touching in Fig. 9.6. The person therefore has an influence on the potential curve. The potential along a line through the arm of the person in Fig. 6.13 is shown in Fig. 9.9. The voltage of an insulated down-conductor that is decisive for dimensioning must therefore be determined from Fig. 9.9 in the case

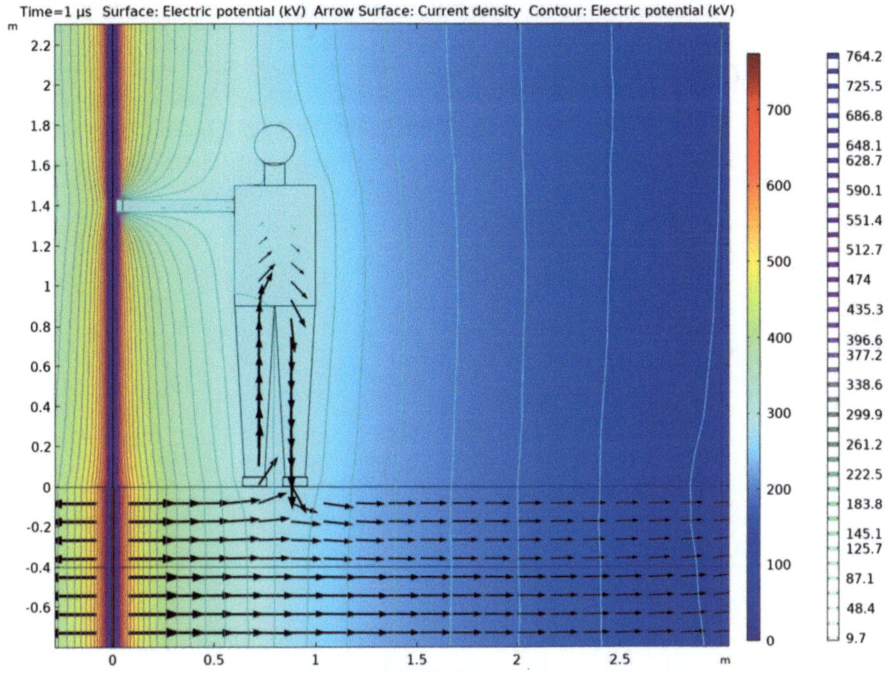

Fig. 9.7 Field for the case that the person does not touch the down conductor. Field figure at the time of the current peak at t = 1 μs with equipotential surface and lines as well as vectors of the current density. Negative subsequent stroke 0.25/100 with a peak value of 24.7 kA according to Table 3.1. Soil: $\rho_E = 250\ \Omega\text{m}$. Current density vectors in logarithmic representation

without contact with the person. In this case, the insulated down-conductor must be dimensioned for a withstand voltage of at least (774 − 280 kV) = 494 kV. This results in the following withstand voltages for an insulated down-conductor in Table 9.5 for the three lightning current shapes and the values according to Table 3.1.

Figure 9.7 shows a current flow through the legs of the person, which is caused by the step voltage. The evaluation of the current height is shown in Fig. 9.10 for the z-component and Fig. 9.11 for the magnitude. The current is galvanically coupled, apart from the capacitively coupled current in the arm, and is proportional to the exciting current 0.25/100. According to the evaluation in Table 9.6, the limit values for the permissible step voltage are clearly exceeded here at a lightning current 10/350.

A person who does not touch the arrester would:

(a) Die from the acting voltage of 494 kV and the associated flashover.
(b) Die from the effect of the step voltage.

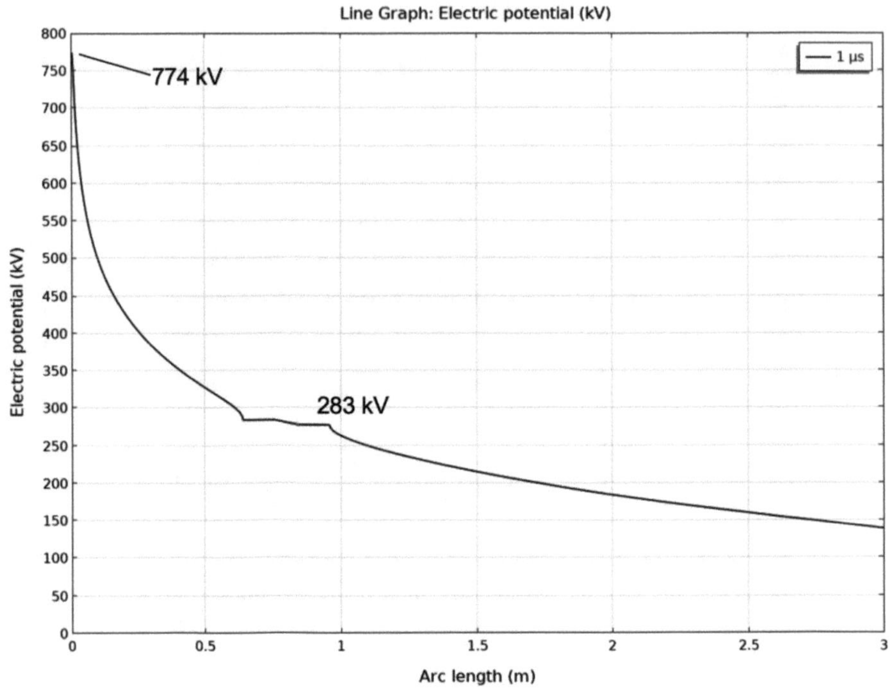

Fig. 9.8 Potential on the earth's surface along line in Fig. 6.9 at the time of the current peak t = 1 µs for 0.25/100

9.2.3 Inductive Coupling U_M

Due to the massive exceeding of the limit values, see Table 9.5, a calculation of the inductive coupling is unnecessary here. However, the inductive coupling is important again for the case of potential control and is therefore only examined for this case in Sect. 9.3.2.

9.3 Effect of Potential Control Measures in Case of a Water-Permeable Layer

A potential equalisation with a grid of 6 m × 6 m with a mesh size of 0.5 m and 6 mm stainless steel rods is located according to Fig. 9.12 at a frost-proof depth of 0.5 m below the person. In the following, the influence of the grid on the galvanic and inductive coupling is calculated with the programme COMSOL-Multiphysics.

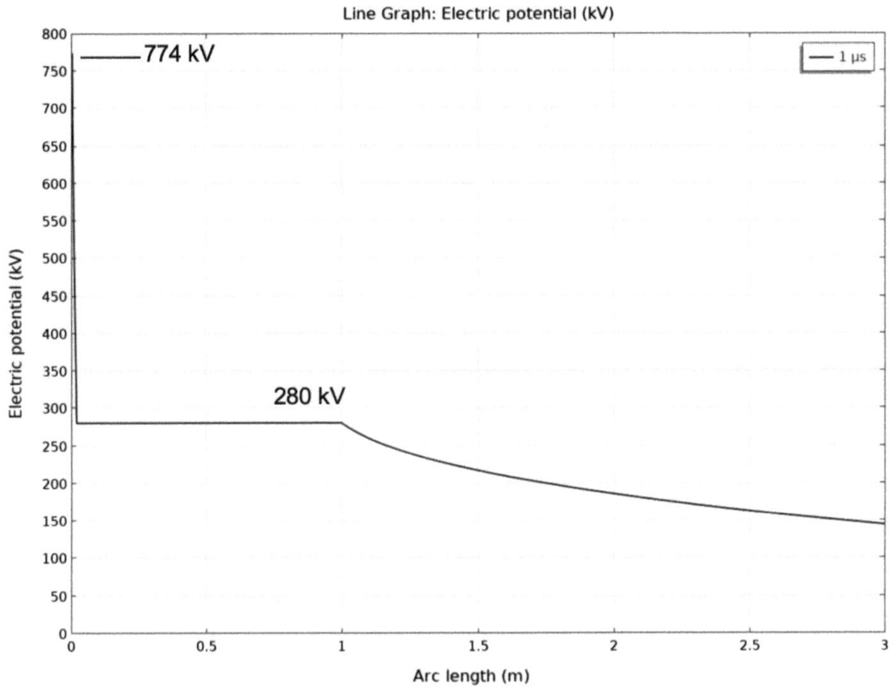

Fig. 9.9 Potential on the earth's surface along line in Fig. 6.13 at the time of the current peak t = 1 μs for 0.25/100

Table 9.5 Results for the case without touching the down conductor. Type A earth electrode; Class I

	U_P	U_S	U_S/U_P
	kV	kV	
10/350	589	140	1.76
1/200	476	840	1.76
0.25/100	280	494	1.76

U_P—Voltage between hand and foot
U_S—Impulse withstand voltage for an insulated down-conductor

9.3.1 Effect on Galvanic Coupling

The field in Fig. 9.13 shows a lower current in the person than in the case without potential control. The potential curve in Fig. 9.14 along the earth's surface is much flatter compared to Fig. 9.6 and the maximum value is, as expected, significantly reduced. The vectors of the current density shown in the person are essentially given by the displacement current of 6.9 A peak value, compare Fig. 7.18. A voltage of 2.83 kV peak drops across the person due to the galvanic coupling and a current of 2.83 A of the form 0.25/100 flows. The resulting energy W and the charge Q are

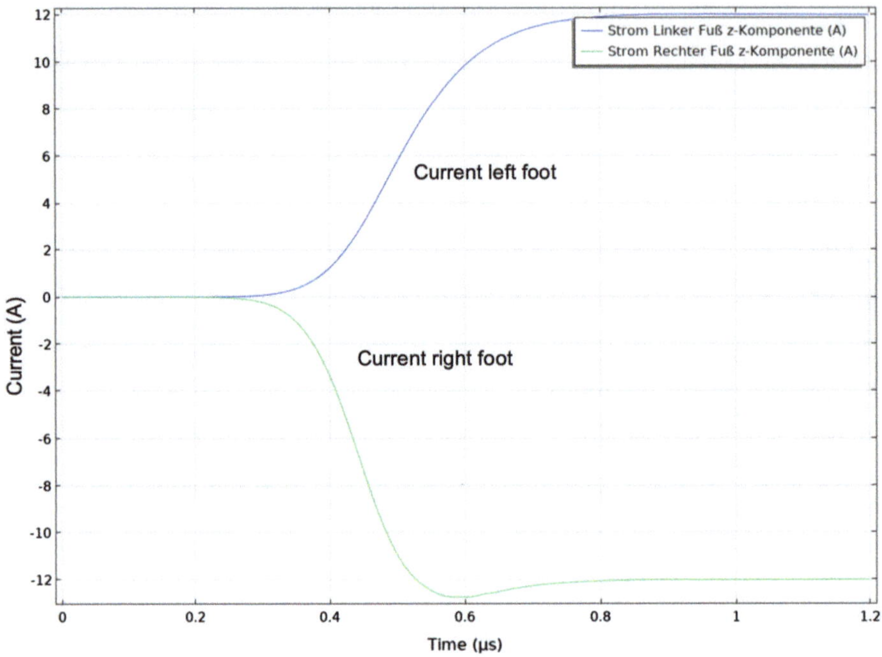

Fig. 9.10 Current curve (z-component) in both legs for 0.25/100. The person does not touch the down conductor

shown in Table 9.7 and exceed the limit values in spite of the potential control. In addition, there is the effect of the induced voltage, which is examined in Sect. 9.3.2.

9.3.2 Effect of Inductive Coupling

The calculation follows the same principle as in Sect. 7.7.2.

The field plot in Fig. 9.15 shows the vectors of the total current density from displacement current and conduction current. The potential curve on the earth's surface in Fig. 9.16 shows a potential of about 70 kV at the feet of the person. However, according to Fig. 9.17, there are high field strength values at the feet of the person and therefore a corona is possible with the negative subsequent stroke. With the negative first flash 1/200 and the positive flash 10/350, there is no danger of gliding discharges because of the lower field strength at the feet. The voltage drop between the feet and the arrestor reduces the current in the person compared to the case when the person is standing on an ideal equipotential surface. The current flow through the person is shown in Fig. 9.18 and reaches a peak value of 157 A, whereby the average pulse width is only 0.15 µs. This results in the energy converted in the person, also shown in Fig. 9.18, with a final value of 2.61 Ws, which exceeds the

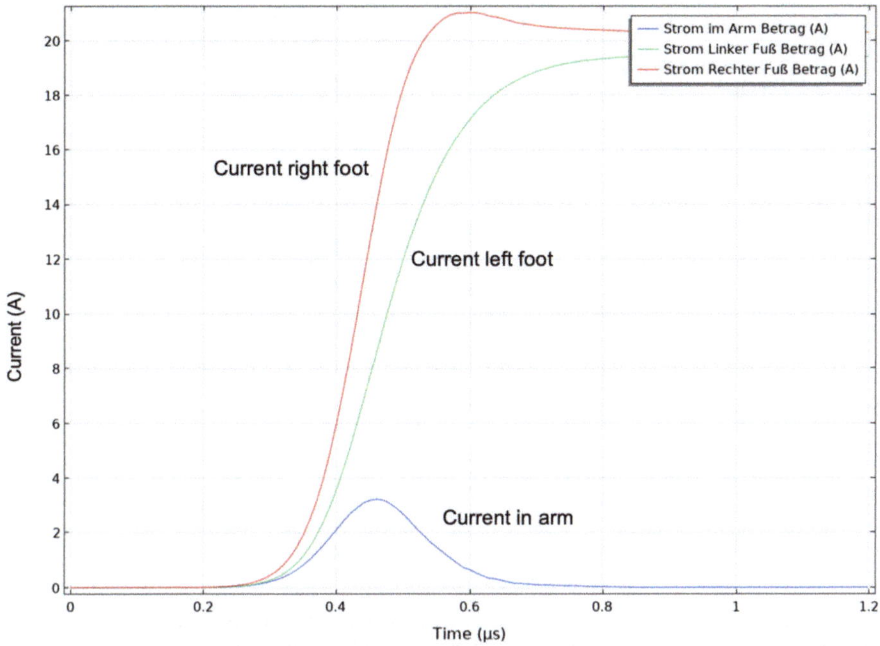

Fig. 9.11 Current curves (magnitude) in both legs for 0.25/100. The person does not touch the down conductor

Table 9.6 Peak value of current I. Earthing system Type A; Class I

	I_B	I	Limit value for step voltage [34]
	kA	A	A
10/350	52	42	25
1/200	42	34	No value defined. The value for 10/350 is used
0.25/100	24.7	20	

I_B—Peak value of the injected lightning current according to Table 3.1
I—Peak value of the current in the body of a person

permissible limit of 1 Ws. The charge is below the limit of 1 mAs. Table 9.8 shows the overview of the results also for the lightning currents 1/200 and 10/350. The negative subsequent stroke is only dangerous in lightning protection classes I and II. The negative first flash is only dangerous in the lightning protection class I. The 10/350 lightning current, on the other hand, is not dangerous. The value for the energy W can be converted quadratically to the other lightning protection classes. However, it must be taken into account that 50% of all negative cloud-to-earth lightning strikes also contain a negative subsequent stroke. Then, according to the current state of science, the energy components have a cumulative effect, so that there is actually a danger in all classes for combined stress. On the other hand, there is a greater danger

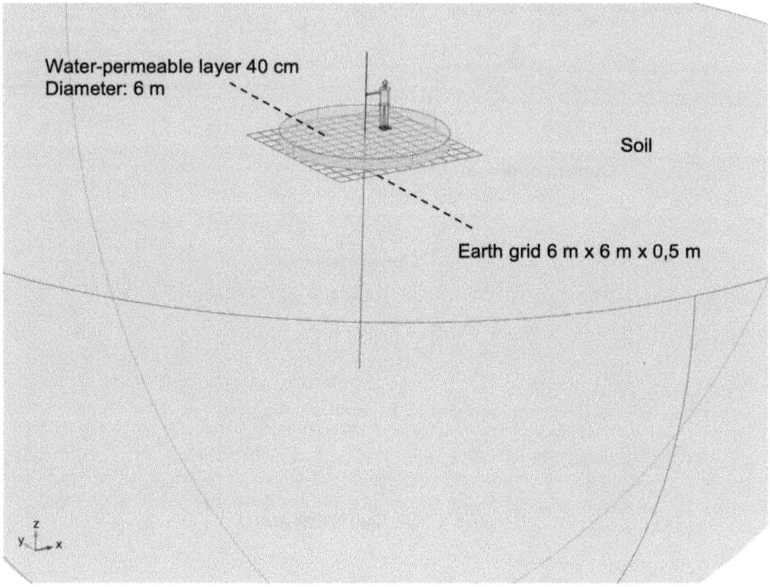

Fig. 9.12 Water-permeable layer with a thickness of 40 cm with equipotential bonding at a depth of 0.5 m consisting of a 6 m × 6 m × 0.5 m mesh grid with 6 mm diameter stainless steel bars

due to the galvanic coupling in Sect. 9.3.1. For physical reasons, both effects, the galvanic coupling and the inductive coupling, act simultaneously, so that for all types of lightning currents and lightning protection classes, the permissible limit values are exceeded and there is a danger of surface discharges and other measures must be taken to solve the problem, which are described in more detail in Chap. 12.

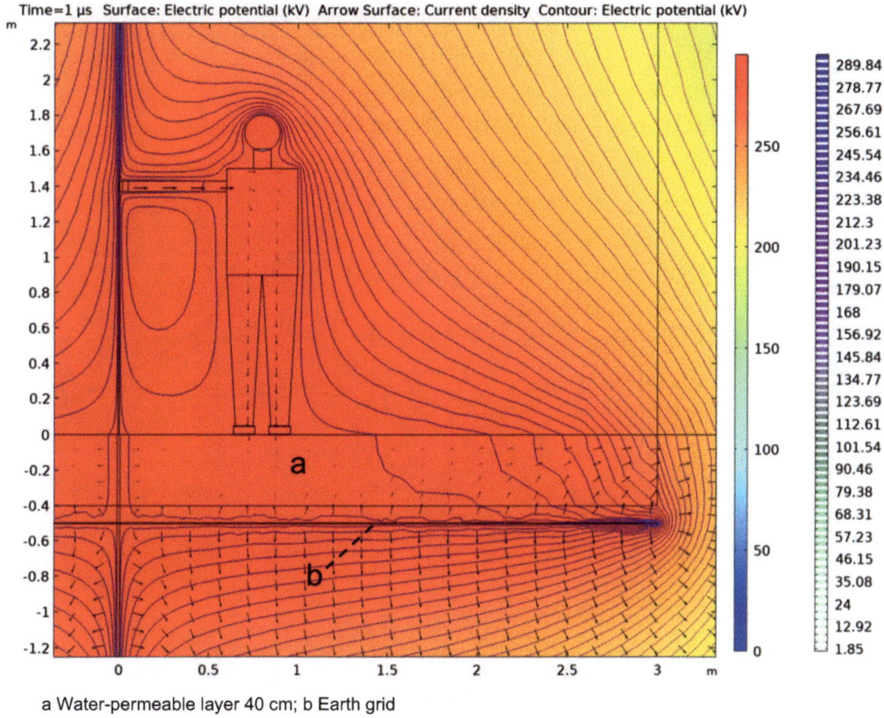

Fig. 9.13 Field at the time of the current peak at t = 1 μs with equipotential surfaces and lines as well as vectors of the current density. Negative subsequent stroke 0.25/100 with a peak value of 24.7 kA according to Table 3.1. Soil ρ_E = 250 Ωm. Current density vectors in logarithmic representation

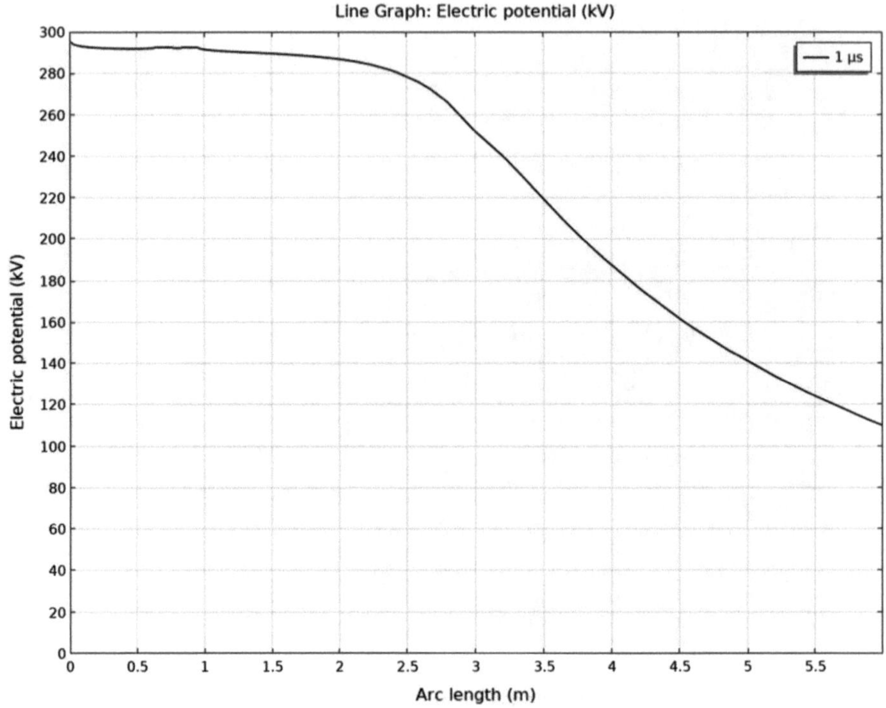

Fig. 9.14 Potential curve on the earth's surface along line in Fig. 6.9 at the time of the current peak t = 1 μs for 0.25/100

Table 9.7 W, Q with water-permeable layer. Earth electrode type A; Class I. Person touched at a height of L = 1.5 m at a distance of 0.8 m, exceeded limit values are shown in bold

	I_B	V_{Hand}	V_{Foot}	U_P	I_K	W	Q
	kA	kV	kV	kV	A	Ws	mAs
+10/350	52	622	616	6	5.96	**9.11 (I)** **5.12 (II)** **2.27 (III–IV)**	**2.9**
−1/200	42	502.3	497	5.3	4.8	**3.33 (I)** **1.87 (II)** **0.83 (III–IV)**	**1.4**
−0.25/100	24.7	295.4	292.6	2.8	2.83	0.57	0.4

I_B—Peak value of the injected lightning current according to Table 3.1
I_K—Peak value of the body current at the time of the current peak
V_{Hand}—Potential at the point of contact of the hand against distant earth
V_{Foot}—Potential at the foot against distant earth
U_P—Voltage between hand and foot

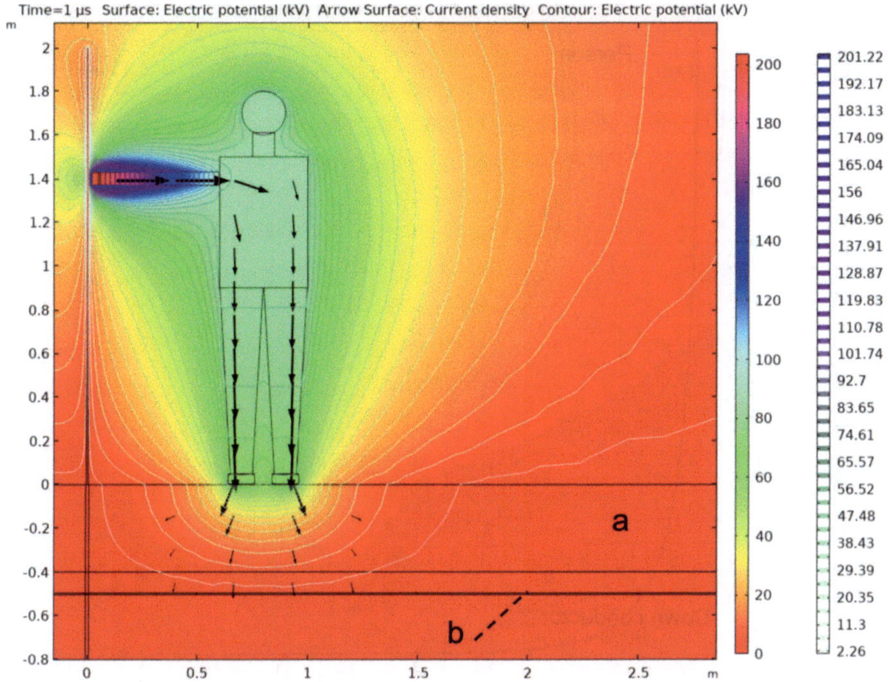

a Water-permeable layer 40 cm; b Earth grid

Fig. 9.15 Field at the time of the maximum current slope at t = 0.45 µs with equipotential surfaces and lines as well as vectors of current density. Negative subsequent stroke 0.25/100 with a peak value of 24.7 kA according to Table 3.1. Water-permeable layer: $\rho_W = 450\ \Omega m$; soil mit $\rho_E = 250\ \Omega m$. Current density vectors in logarithmic representation

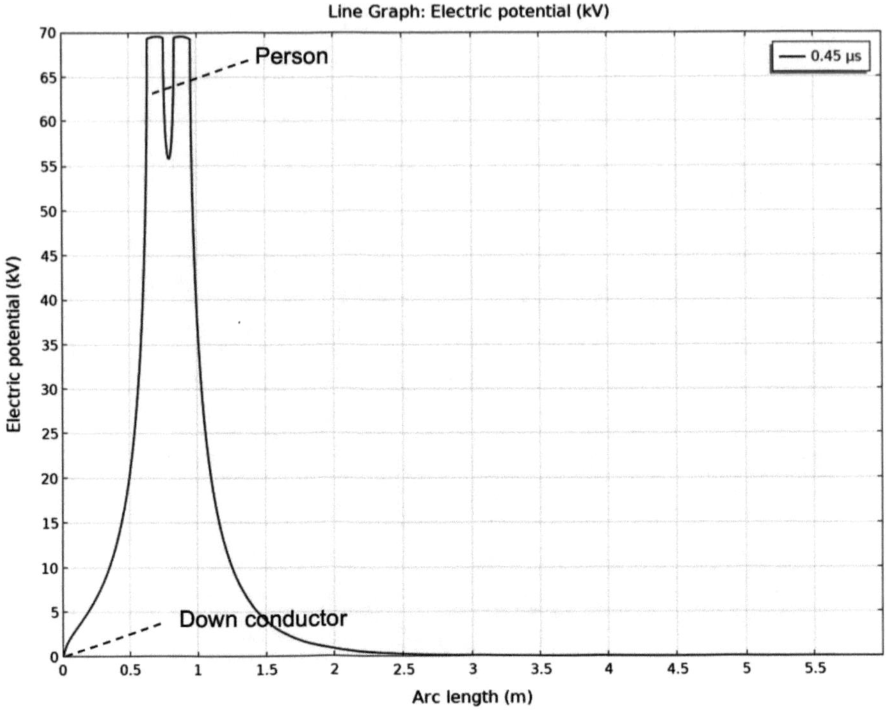

Fig. 9.16 Potential on the earth's surface along line in Fig. 6.9 at the time of the maximum current slope t = 0.45 µs for 0.25/100

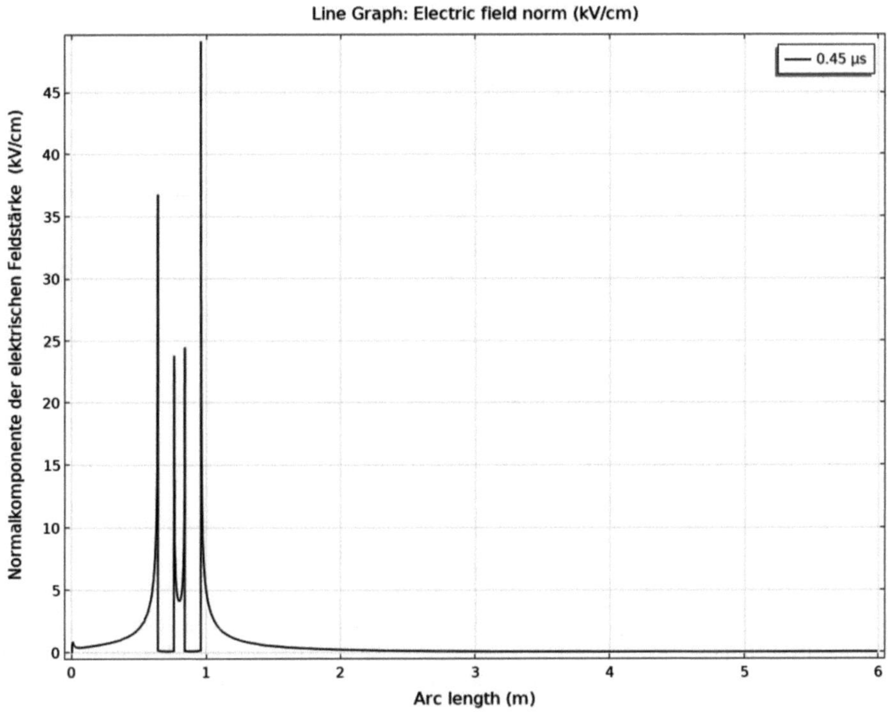

Fig. 9.17 Course of the electric field strength on the earth's surface along line in Fig. 6.9 at the time of the maximum current gradient t = 0.45 μs for 0.25/100

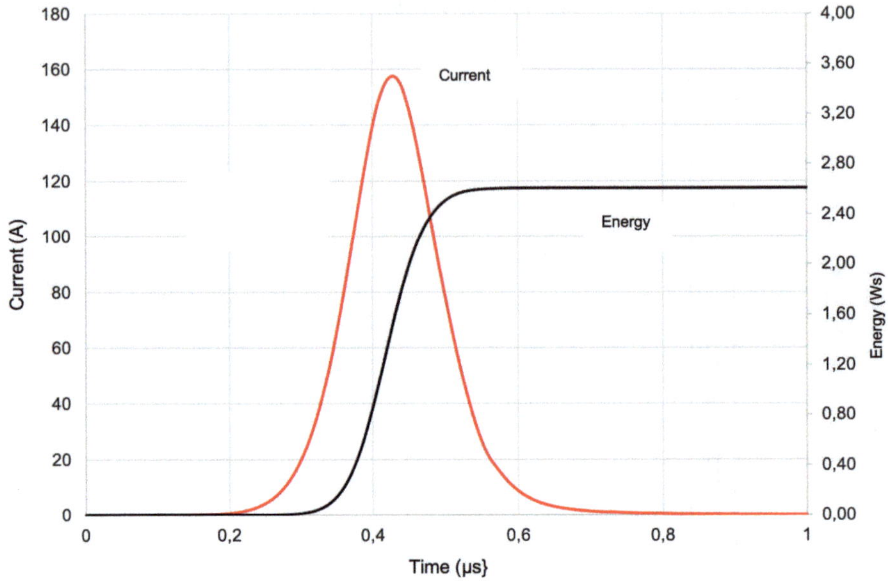

Fig. 9.18 Current in the person and energy W converted in the person by the inductive effect of the injected lightning current. Negative subsequent stroke 0.25/100 with a peak value of 24.7 kA according to Table 3.1. Water-permeable layer: $\rho_W = 450\ \Omega m$; soil mit $\rho_E = 250\ \Omega m$

Table 9.8 Results from the effect of inductive coupling with a water-permeable layer and potential equalisation. Type A earth electrode; Class I. Values for other lightning protection classes in brackets

	I_B	U_{ind}	V_{Hand}	V_{Foot}	I_{KM}	W from I_{KM}	Q from I_{KM}
	kA	kV	kV	kV	A	Ws	mAs
+10/350	52	10.58	10.3	3.9	7.03	0.25	0.053
−1/200	42	86.9	86.9[a]	33.2	58.4	**1.65 (I)** 0.92 (II) 0.41 (III–IV)	0.042
−0.25/100	24.7	203	203[a]	70	157	**2.61 (I)** **1.46 (II)** 0.65 (III–IV)	0.047

I_B—Peak value of the injected lightning current according to Table 3.1
I_{KM}—Peak value of the body current at the time of the current peak due to magnetic coupling
V_{Hand}—Potential at the point of contact of the hand against distant earth
V_{Foot}—Potential at the foot against distant earth
U_{ind}—Voltage between hand and foot
[a] The difference between potential at the hand and at the foot is also influenced by the capacitive displacement component. Therefore, the voltage drop at the person is not determined by Ohm's law alone

Chapter 10
Danger Due to Step Voltage

In addition to the touch voltage, the step voltage must also be calculated and evaluated. The step voltage has already appeared as a side effect in the previous simulations of the touch voltage in some arrangements (especially without touching the person at the down conductor).

Also, the probability of a person being endangered by step voltage is greater, since the step voltage acts in the entire area of the voltage funnel in the vicinity of a down-conductor inserted into the earth. The statistics of personal injuries in Chap. 14 also show that about 50% of all personal injuries were due to step voltage.

For the calculation of step voltage, inductive coupling does not have to be taken into account because of the higher limit value. Therefore, programs available on the market, e.g. XGS-Lab [13] can be used, which are mainly used for the design of 50/60 Hz earthing systems, and which allow a comfortable visualisation of the results in accordance with the requirements of standards. The programs provide both the touch voltage and the step voltage as a result, whereby the inductive coupling between the down conductor and the surface covered by the touching person is not taken into account.

An introduction to the calculation of step voltage is presented in detail in Chap. 21 and provides an introduction to understand the following.

In the following sections, the two sample buildings shown in Chap. 3 with earth electrode types A and B are calculated for the values according to Table 3.1.

10.1 Earthing System Type A

The earth electrode type A is described in detail in Chap. 3. The evaluation in XGS-LAB now allows different representations. In the following representations, Figs. 10.1, 10.2 and 10.3, the limit values of 25 kV 10/350 for step voltage and 2 kV for touch voltage apply.

J. Meppelink, *The Lightning Rod as a Danger*,
https://doi.org/10.1007/978-3-031-30434-7_10

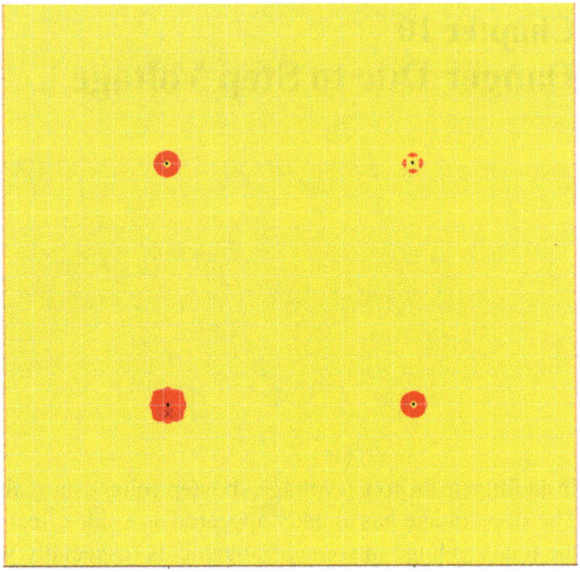

Fig. 10.1 All references 0.25/100 1 MHz 17730 A eff. Grid size: 1 m

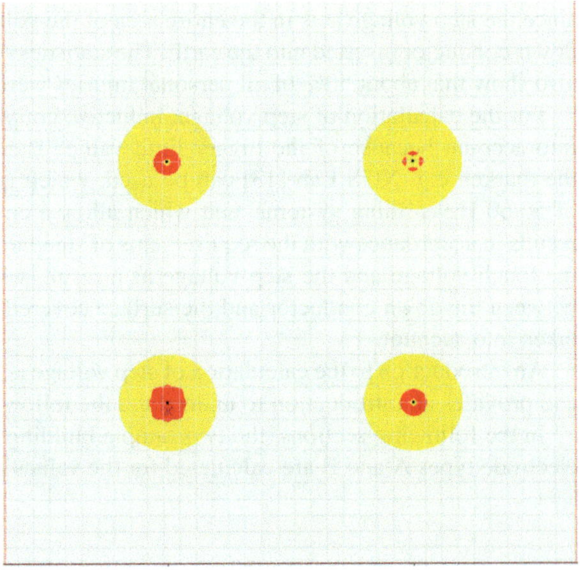

Fig. 10.2 Reach distance 3 m 0.25/100 1 MHz 17730 A eff. Grid size: 1 m

– All references, Fig. 10.1: This means that a person standing in the yellow area and touching a down conductor is in danger. **Red means that the step voltage there is dangerous**.

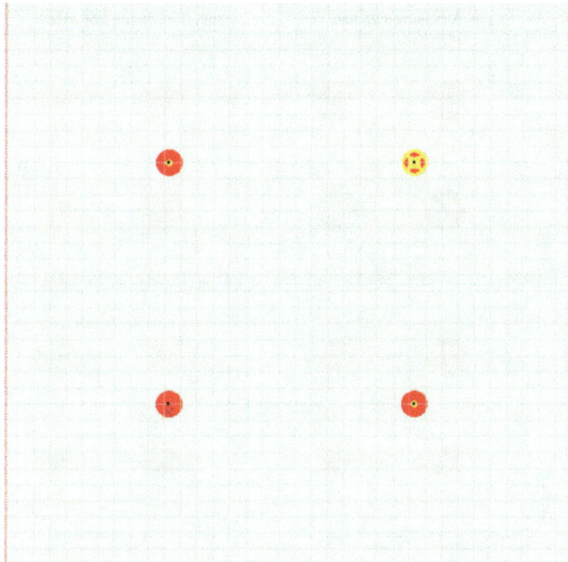

Fig. 10.3 Reach distance 0.8 m 0.25/100 1 MHz 17730 A eff. Grid size: 1 m

– Reach distance 3 m, Fig. 10.2: This means that a person standing in the yellow area and touching the arrester is in danger. The 3 m distance was chosen because of IEC 62305-3. **Red again means: step voltage is dangerous.**
– Reach distance 0.8 m, Fig. 10.3: This means that a person touching the arrester from a distance of 0.8 m is also at risk. **Red is then to be understood as meaning that both the step voltage and the touch voltage are dangerous.**

The Fig. 10.4 shows a comparative calculation for all three types of lightning currents.

The hazardous areas for touch and step voltage are according to the colour coding shown below and apply to all following figures.

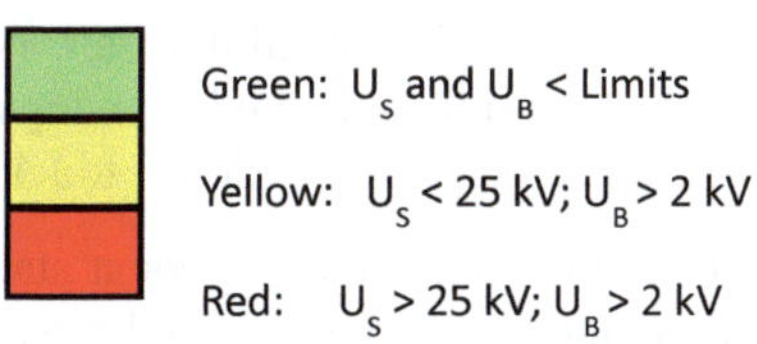

Note What is the meaning of all references? The touch voltage is always given with reference to the distance of the person to the down conductor. All references means that a person stands within the entire yellow area and touches the down conductor from there, also indirectly. In the other cases, the reach distance is given, which is more in line with practice.

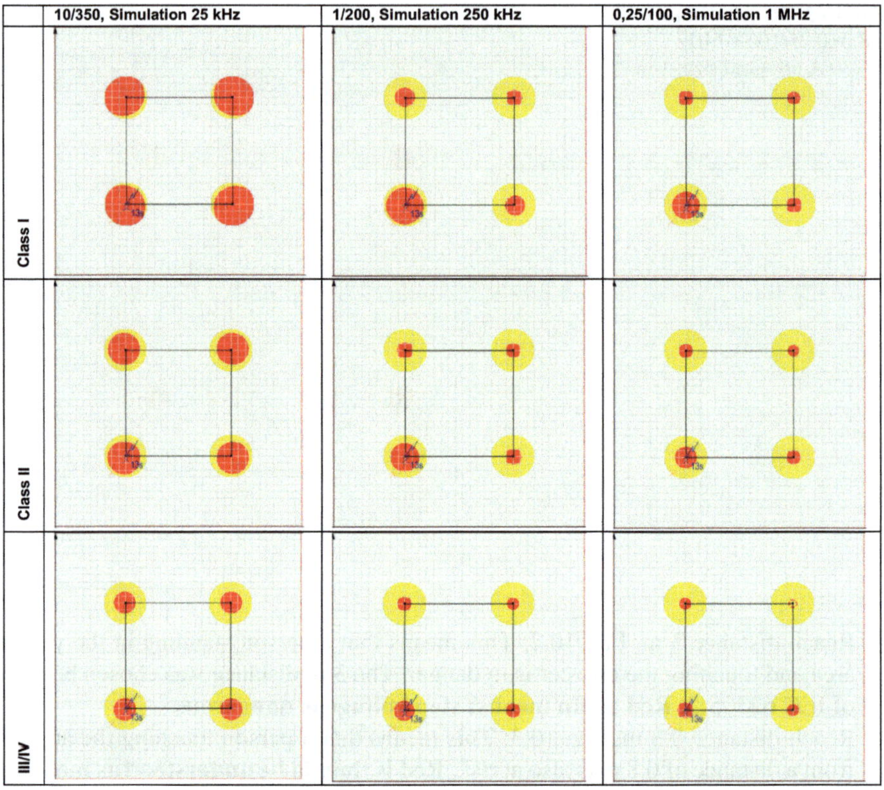

Fig. 10.4 Comparative analysis of step and touch voltage with XGS-LAB for all lightning current forms and classes for earth electrode type A. Soil: $\rho_E = 250\ \Omega\text{m}$. **Reach distance 3 m**. Grid size: 1 m

10.2 Earthing System Type B

The earth electrode type B is described in detail in Chap. 3. The evaluation in XGS-LAB now allows different representations. In the following representations, the limit values of 25 kV 10/350 for step voltage and 2 kV for touch voltage apply.

- All references, Fig. 10.5. This means that a person standing in the yellow area and touching a down conductor is in danger. **Red means that the step voltage there is dangerous**.
- Reach distance 3 m, Fig. 10.6: This means that a person standing in the yellow area and touching the arrester is in danger. The 3 m distance was chosen because of IEC 62305-3. **Red again means: step voltage is dangerous**.
- Reach distance 0.8 m, Fig. 10.7: This means that a person touching the arrester from a distance of 0.8 m is also at risk. **Red is then to be understood as meaning that both the step voltage and the touch voltage are dangerous**.

Fig. 10.5 All References 0.25/100 17730 A. Grid size: 1 m

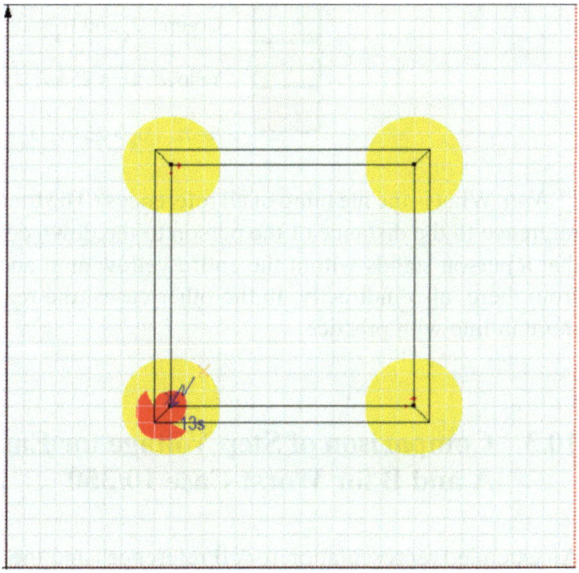

Fig. 10.6 Reach distance 3 m. Grid size: 1 m

– The Fig. 10.8 shows a comparative calculation for all three types of lightning currents.

The hazardous areas for touch and step voltage are according to the colour coding shown below and apply to all following figures.

Fig. 10.7 Reach distance 0.8 m. Grid size

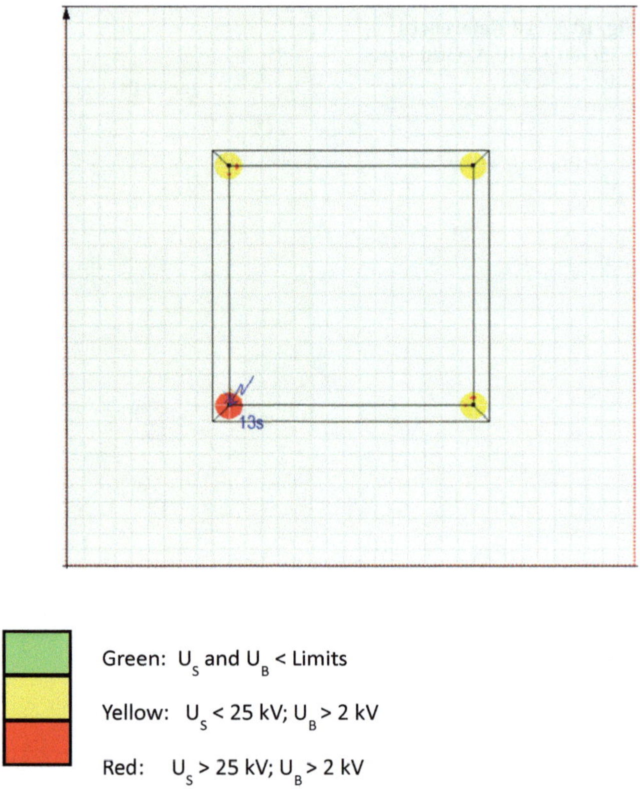

Note What is the meaning of all references? The touch voltage is always given with reference to the distance of the person to the down conductor. All references means that a person stands within the entire yellow area and touches the down conductor from there, also indirectly. In the other cases, the reach distance is given, which is more in line with practice.

10.3 Comparison of Step Voltage for Earth System Type A and B for Worst Case 10/350

An extended representation for all references and the worst case 10/350 is shown in Fig. 10.9. In this comparison it becomes clear that the step voltage for earth electrode type B is worse, i.e. for the same load with 100 kA, the range of dangerous step voltage is greatest there.

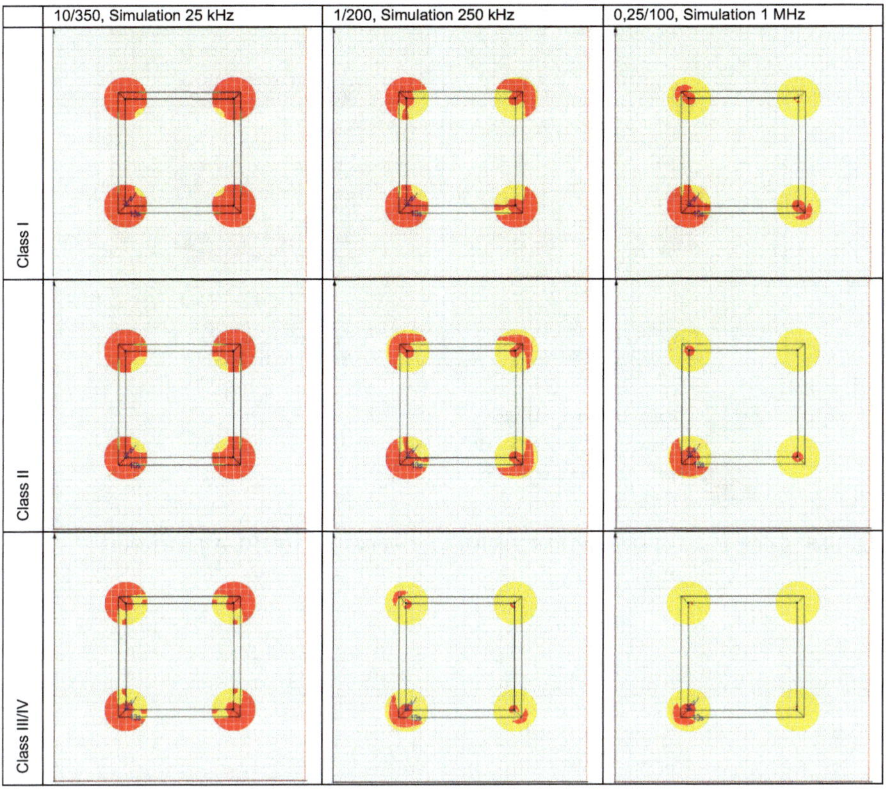

Fig. 10.8 Total step and touch voltage analysis with XGS-LAB for all lightning current forms and classes for earth electrode type B. Soil: $\rho_E = 250\ \Omega m$. **Reach distance 3 m**. Grid size: 1 m

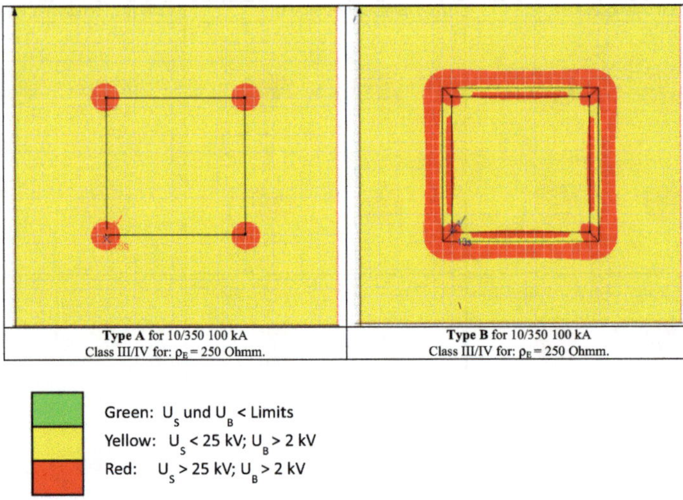

Fig. 10.9 Comparison of step and touch voltage for worst case 10/350. All references

Chapter 11
Summary of the Hazard Posed by a Lightning Rod

The following results show the physical effects and their impact on people. The probability of contact is commented on in Chaps. 15 and 16.

11.1 Danger from Touch Voltage When Touching a Bare Down Conductor

The following summary in Table 11.1 concerns the case that occur when a person touches a down conductor. Values apply to specifications in Table 3.1 and the lightning current forms for calculations (Heidler functions) specified in IEC 62305-1, Appendix B. In the overall result, the decisive factor is whether the hazard due to A (10/350) or B (1/200) predominates. A negative subsequent stroke cannot occur alone without a negative first flash. Therefore, a green field in area C with a red area B is still fatal.

11.2 Danger from Proximity and Step Voltage Without Touching a Bare Down Conductor

The electric field coupling by proximity U_N is shown in Chap. 6.4 and Figs. 6.28, 6.29 and 6.30. This results in the areas in Table 11.2 where a person would be killed by an electric flashover of the air.

J. Meppelink, *The Lightning Rod as a Danger*,
https://doi.org/10.1007/978-3-031-30434-7_11

Table 11.1 Danger due to touch voltage. Earth electrode type A with $\rho_E = 250\ \Omega m$. Bare conductor. Red: fatal consequence due to ventricular fibrillation. Green: survival

Insulation above soil	Condition	Potential control measures	Coupling mechanism	Danger: A:10/350 B:1/200 C:0,25/100	I	II	III/IV	Overall result
None (Soil only)	D	None	Galvanic	ABC				Death
			Inductive	ABC			(2)	
		Earth grid	Galvanic	A				
				B				
				C				
			Inductive	C			(2)	
Asphalt wet 5 cm	D/W	None	Galvanic Surface discharge	ABC				Death
		Earth grid	Inductiv	A				
				B				
				C				
Asphalt and 2 cm Water		None	Galvanic	ABC				Death
		Earth grid	Galvanic	A				
				B				
				C				
			Inductive	A				
				B				
				C				
Water-permeable layer	D (1)							
	W	None	Galvanic	ABC				Death
		Earth grid	Galvanic	AB				
				C				
			Inductive	A				
				B				
				C				

D—Dry
W—Wet
(1) A dry water-permeable layer behaves like dry asphalt
(2) Using the current slope values specified in IEC 62305-1, Table 6.1 (100 kA/μs for lightning protection class III/IV), this area becomes harmless in terms of induced voltage. Danger from proximity and step voltage without touching a bare down conductor

Table 11.2 Danger from proximity flashover and step voltage. Earth electrode type A with $\rho_E = 250\ \Omega m$. Bare down-conductor. Red: fatality due to ventricular fibrillation. Green: survival

Insulation above soil	Condition	Potential control measures	Coupling mechanism	Danger A:10/350 B:1/200 C:0,25/100	I	II	III/ IV	Overall result
None (Soil only)	D	None	U_N causes flashover only at distances as per Figs. 6.28, 6.29 and 6.30	AB	(1)	(2)	(3)	Death
			Step voltage					
		Earth grid	No danger for U_N and step voltage					
Asphalt wet 5 cm	D/W	None	U_N causes flashover only at distances as per Figs. 6.28, 6.29 and 6.30	AB	(1)	(2)	(3)	Death
		Earth grid	No danger for U_N and step voltage	ABC				
			Inductive Flashover only at distances as per Figs. 6.28, 6.29 and 6.30	C	(4)	(4)		Death Class I, II
Asphalt and 2 cm Water		None	U_N causes flashover only at distances as per Figs. 6.28, 6.29 and 6.30 Step voltage	ABC	(1)	(2)	(3)	Death
		Earth grid	No danger for U_N and step voltage	ABC				
			Inductive Flashover only at distances as per Fig. 17.1	A				Class I,II
				B	(4)			
				C	(4)	(4)		
Water-permeable layer	D (5)							
	W	None	U_N causes flashover only at distances as per Figs. 6.28, 6.29 and 6.30 Step voltage	ABC				Death
		Earth grid	Inductive Flashover only at distances as per Fig. 17.1	A				
				B	(4)			Death Class I
				C	(4)	(4)		Death Class I,II

(1) Section 6.4 und Fig. 6.28
(2) Section 6.4 und Fig. 6.29
(3) Section 6.4 und Fig. 6.30
(4) Only in areas of danger as per Fig. 17.1
(5) A dry water-permeable layer behaves as a dry layer of asphalt
D—Dry
W—Wet

Chapter 12
Measures to Reduce Step and Touch Voltage as Per IEC 62305-3

In Chap. 8 of standard IEC 62305-3, normative measures are described. However, these measures obviously refer to dry conditions. Therefore, a wet surface was also taken into account in this study. In the following, the individual measures are compared with the results of this study and commented on.

12.1 Distance

The standard [12] requires a distance of 3 m from down conductors. This means that a person must not touch the down conductor. This distance of 3 m is also correct according to the calculations for a specific earth resistance of 250 Ωm, see Figs. 6.28, 6.29 and 6.30. This applies to the case of electrical field coupling U_N.

If, on the other hand, potential control is used, the coupling is reduced to inductive coupling and therefore the danger zones (person not touching the metal down conductor) must be reassessed. A detailed calculation is given in Chap. 16. Figure 17.9 shows the danger zones with inductive coupling and metallic down conductor. It turns out that the hazard area shrinks to 5 cm for lightning protection class I. The hazard area for lightning protection class III is reduced to 5 cm. In lightning protection class III/IV, the danger zone is reduced to about 1 cm.

As can also be seen in the analysis of accident statistics in Chap. 14, fatal accidents practically do not occur due to direct contact. It then remains to assess the probability of a person in a thunderstorm incident:

1. stands in front of a down conductor within the critical distance of 5 cm (lightning protection Class I) (flashover resulting in death).
2. touches it (fatality).

J. Meppelink, *The Lightning Rod as a Danger*,
https://doi.org/10.1007/978-3-031-30434-7_12

12.2 Site Insulation

IEC 62305-3 requires a 5 cm layer of asphalt or a 15 cm layer of gravel. The electrical strength of dry asphalt is given in [28] as 10–60 kV/mm. With a thickness of 5 cm, breakdown is not to be expected. However, surface discharges do occur, which are not mentioned in the standard. If there is a layer of water on an insulating layer, the exposure of a person touching the arrester is too high because of the galvanic coupling. Equipotential bonding is also not effective. Here, an insulating down conductor is a possible solution.

An improvement of the IEC 62305-3 standard is not expected until the next ED 4, which will probably not be published until 2030.

12.3 Signage

Signage indicating to keep a distance from the diversion is not wrong. However, there is criticism of this because these signs are not seen in the rain and in the dark and these signs would have to be in all languages. Alternatively, a pictogram would be a solution.

12.4 Insulation of the Down Conductor

IEC 62305-3 requires insulation of the down conductor with a layer of polyethylene at least 3 mm thick. A withstand impulse voltage of 100 kV 1.2/50 is required. It is not specified how this impulse withstand voltage is to be determined. The standard does not refer to the reduction of the electrical strength due to surface discharges. The specified value of 100 kV is unrealistic, as the previous calculations show. These discrepancies must also be eliminated in ED 4 of IEC 62305-3.

Chapter 13
Insulating Down Conductor

13.1 Requirements for Insulating Down Conductor

As according to the state of the art, a high-voltage-resistant insulated down conductor is used as the insulation of a down-conductor. These were originally used with the aim of reducing the separation distance.

Due to the current discussion about the danger of touch voltage, insulated down conductors are proposed as a solution to the problem [17]. The following solutions can be considered:

(a) Use of a high-voltage-resistant insulated down conductor in the area where a person can touch the down conductor or is standing close to a down conductor.
(b) Use of a high-voltage-resistant insulated down conductor, which is used as a complete down conductor on the building due to the requirement of a smaller separation distance.

For both variants (a) and (b), the withstand voltage U_S of a high-voltage-resistant insulated down conductor must be determined. This determination was investigated in detail in the previous sections. The resulting values are summarised in Table 31. It can be seen that in the case of galvanic coupling without equipotential bonding, very high withstand voltages would have to be fulfilled and a distinction must be made between the costs for equipotential bonding and the high-voltage-resistant insulated down-conductor. Therefore, in the following, the high-voltage-resistant insulated down-conductor is only examined for the case of existing equipotential bonding. All calculations are then limited to the case of inductive coupling (Table 13.1).

J. Meppelink, *The Lightning Rod as a Danger*,
https://doi.org/10.1007/978-3-031-30434-7_13

Table 13.1 Necessary withstand voltages for insulated down-conductor for loads according to Table 1.1, earth electrode type A

Insulation above soil	Con-dition	PM	Cou-pling	Shape	Minimum withstand voltage U_S for insulated down-conductor kV								
					10/350			1/200			0,25/100		
					I	II	II-IV	I	II	II-IV	I	II	II-IV
None, see 6.2.2	D	None	Gal-vanic	i(t)	1000	750	500	814	610	407	482	362	241
		Grid	Induc-tive	di(t)/dt	11,4	8,55	5,7	88,7	66,5	44,4	205	154	103
Asphalt 5 cm	D/W	None	Gal-vanic	i(t)	1013	759	507	818	613	409	481	361	240
		Grid	Induc-tive	di(t)/dt	11,4	8,55	5,7	88,7	66,5	44,4	205	154	103
Asphalt and 2 cm Water		None	Gal-vanic	I(t)	1004	753	502	809	607	405	484	363	242
		Grid	Induc-tive	di(t)/dt	6,25	4,7	3,12	57,5	43,1	28,8	150,4	112,8	75,2
Water-per-meable layer	D 1)	None	Gal-vanic	i(t)	1040	780	520	840	630	420	494	371	247
		Grid	Induc-tive	di(t)/dt	11,4	8,55	5,7	88,7	66,5	44,4	205	154	103
	W	None	Gal-vanic	i(t)	1027	770	514	810	608	405	494	371	247
		Grid	Induc-tive	di(t)/dt	11,4	8,55	5,7	88,7	66,5	44,4	205	154	103

A dry water-permeable layer behaves as a dry layer of asphalt
Red areas: Use of insulating down conductors is not recommended sue to economic reasons
Grey areas: Products (insulating down conductors) are available on market
D: Dry
W: Wet
PM: Potential control measures

13.2 Electric Field Strength Along an Insulating Down Conductor

An insulating down conductor is intended to protect people who touch it. The worst-case scenario is that a person touches the insulated down conductor with a finger while the arm is stretched out, thus maximising the surface area of the down conductor. It is practically impossible for a person to fully enclose an insulated down conductor with their hand, as insulated down conductors are mounted close to a wall.

The electric field strength in the insulating material and in the air space in front of the insulated down conductor is decisive for the dimensioning of an insulated down conductor for the protection of persons. Contact has an influence on the electric field strength in the insulating material. In the case of contact, it must also be investigated whether a surface discharge occurs on the surface of the insulating material, which would destroy the insulation in the event of a surface discharge flashover.

Furthermore, it must be investigated whether a flashover from the surface of the insulated down conductor to a person standing next to it can occur even without a person touching the insulated down conductor and whether the insulated down conductor can be bridged by a surface discharge and thus become a danger for the person. Therefore, two simulations are carried out:

(a) 3-D simulation to determine the conditions when the insulated down-conductor is touched in order to determine the coupled current.
(b) 2-D simulation to determine the electric field strength in the space between the insulated down conductor and a person.

13.2.1 Electrical Stress When a Person is in Contact with an Insulating Down Conductor

A 3-D field calculation is carried out for the arrangement according to Fig. 13.1. A reasonable simulation must be carried out by touching a person on the surface of the isolated arrester. As a compromise between reality and simulation effort, a point contact is investigated. The touch is realised by a small sphere. The voltage effective during inductive coupling according to Fig. 19.4 with a peak value of 211.7 kV (corresponds to lightning protection class I) is applied as the exciting voltage to the down conductor, which is insulated at the bottom from the earthed plate. From the simulation, the electrical load on the person is determined. The simulation then also shows the load within the insulated down-conductor as a result of the contact. The calculated current waveform when touching an insulated lead is shown in Fig. 13.2 and corresponds to the time derivative of the voltage in Fig. 19.4, which corresponds to a purely capacitive current. The field plot in Fig. 13.3 shows the current density vectors at two different times according to Fig. 13.2, where the direction of the current reverses in the person. The field figure with electric field strength vectors is shown in Fig. 13.4.

An evaluation of the field strength magnitude within and on the surface of the insulating down conductor is calculated in a further refined, purely electrostatic 3-D model according to Fig. 13.5. Here, the insulating down conductor is at 205 kV potential and the person touching it is at earth potential. The field strength in the insulating material and on the surface of the insulating material is determined. The field plot is evaluated in two sectional planes according to Fig. 13.6. There is an increase in field strength in the area between the conductor and the person touching it. The normal component of the electric field strength, however, remains far below the permissible field strength for XLPE according to Fig. 13.9. The field strength on the surface of the insulated arrester is shown in Fig. 13.8. As expected, there is a very high electric field strength in the area of contact with the sphere. As a result, a surface discharge occurs on the surface of the insulating material in this case. Whether a surface discharge flashover occurs between the point of contact and the end of the insulation is a question of the actual current load (lightning protection class), the geometry of the insulated down conductor and the insulating material, see Sect. 20, and can only be checked by a type test (Fig. 13.7).

A person touching the down conductor (worst case with an outstretched arm) is therefore not critical from the point of view of the field strength stress in the XLPE

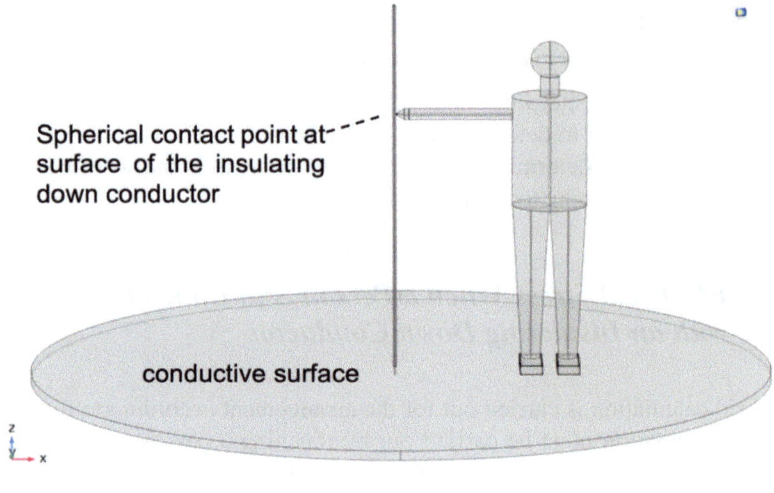

3-D-view

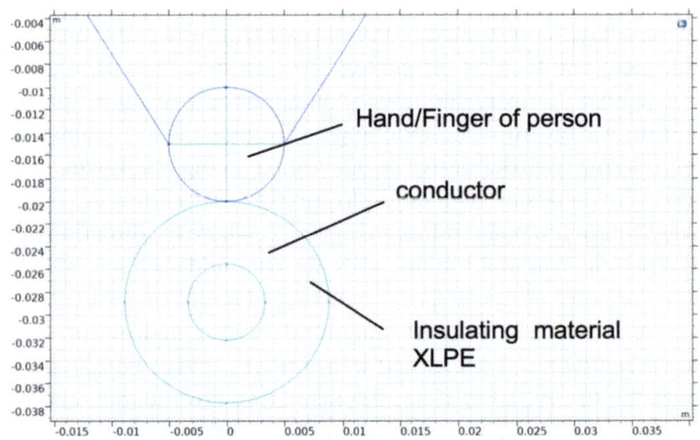

y-x-plane in a height of 1,4 m (Mid of arm of person)

Fig. 13.1 Model for calculating the current density and electric field strength when a person touches a high-voltage insulated conductor. The contact area is shown in higher resolution in Fig. 13.5

insulating material. Depending on the lightning protection class, a surface discharge can occur on the surface. The person is not dangerously exposed to the capacitive current.

An insulating down conductor for the protection of persons must therefore be tested in the laboratory. The standard, IEC 62305–3 does not currently specify this. For a type test, it must be determined.

- How is the contact to be verified?
- What voltage form is to be used for testing?

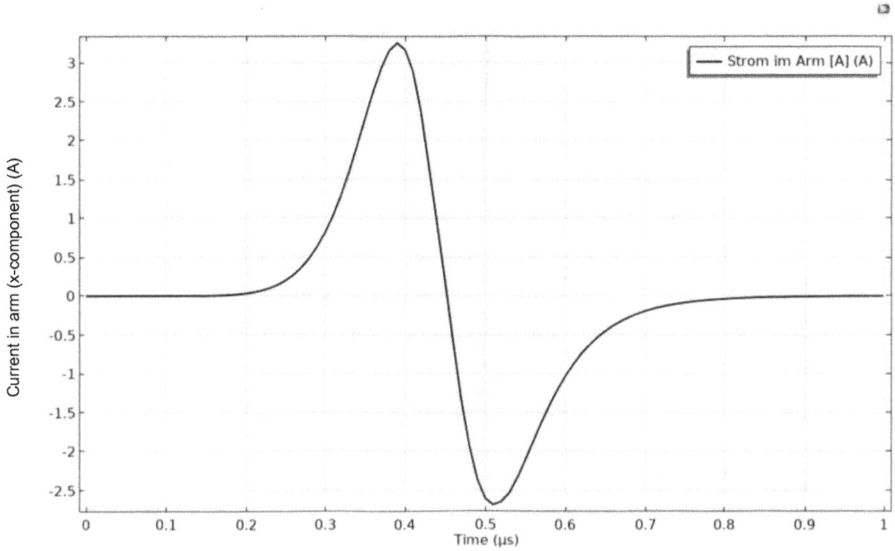

Fig. 13.2 Current in the arm of the person due to capacitive coupling to the insulated down-conductor. The voltage at the down conductor is as shown in Fig. 19.3, (differentiated negative sequence lightning stroke 0.25/100 with a peak value of 24.7 kA multiplied by 1.48 μH) according to Table 13.1. Current density vectors in logarithmic representation

- How can the test voltage be generated and measured in the laboratory?
- When is the type test considered to have been passed?

13.2.2 Electrical Stress When a Person is not Contact with an Insulating Down Conductor

The most probable case in practice is not the contact of the isolated down conductor with the finger but the lateral approach with the body of the person. The resulting conditions are calculated as follows. First, the occurring voltages must be determined as a function of the distance, since the mutual inductance determining the induced voltages depends on the distance s and the height. The calculations are made with the values in Table 13.2 for a height of 1.4 m. The numerical field calculation is carried out as a 2-D simulation according to Fig. 13.12 with the distances s and the occurring induced voltages according to Table 13.2 in x–y coordinates. Figure 13.13 shows two field figures with the distances s = 80 cm and s = 5 cm. The normal component of the electric field strength is shown in Fig. 13.14 for two distances. The evaluation of these calculations shows in Fig. 13.11 that the electric field strength is smaller at small distances s than at large distances s. The electric field strength reaches values of up to 99 kV/cm at large distances. However, since these field strengths occur on the surface of the insulating material at the interface with the surrounding air,

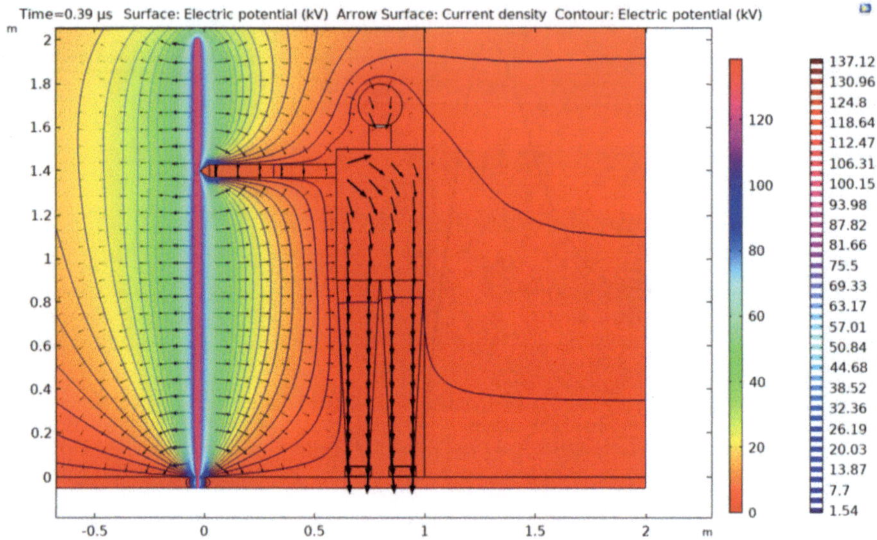

t= 0,39 µs

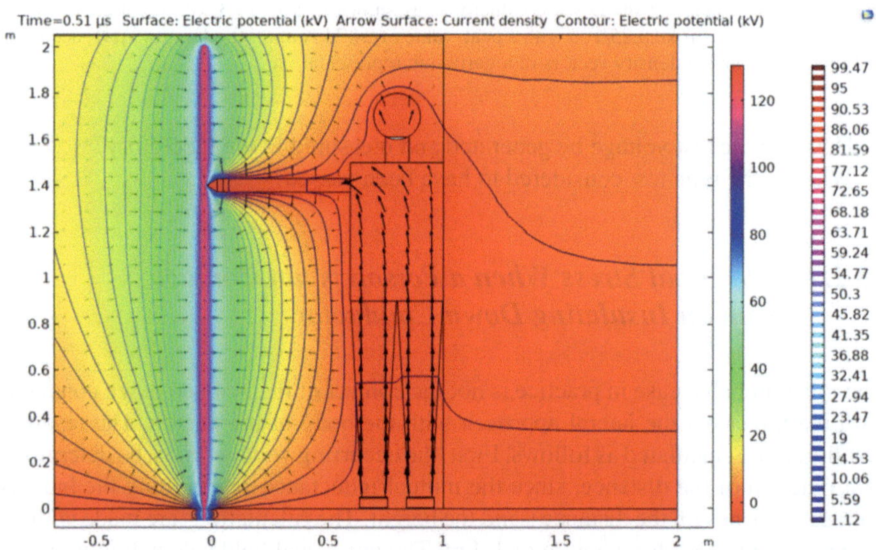

t=0,51 µs

Fig. 13.3 Field at the time of the maximum current slope at t = 0.39 µs and t = 0.51 µs, compare Fig. 13.2 with equipotential surface and lines as well as **current density vectors**. The voltage is at the down conductor according to Fig. 19.3, Current density vectors in logarithmic representation

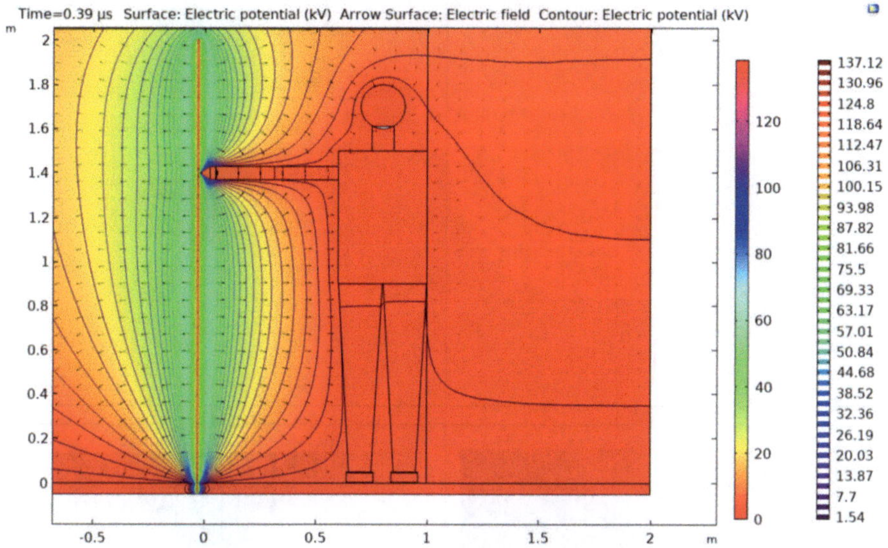

Fig. 13.4 Field at the time of the maximum current slope at t = 0.39 μs with equipotential surfaces and lines as well as vectors of **the electric field strength**. The voltage is at the down conductor according to Fig. 19.3, field strength vectors in logarithmic representation

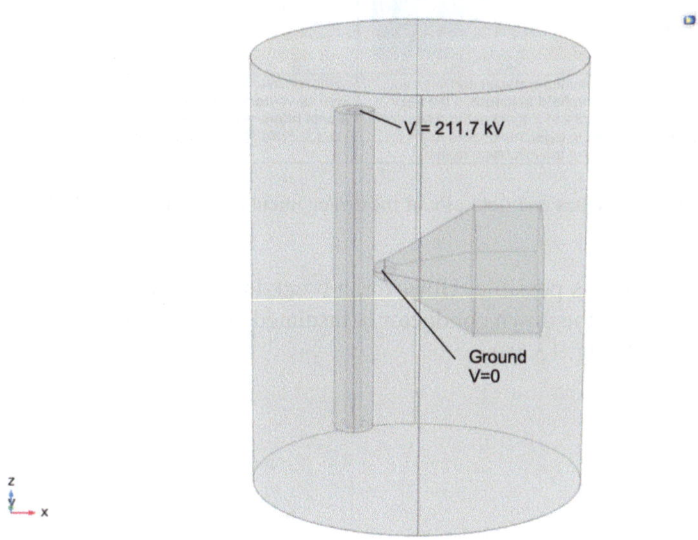

Fig. 13.5 Simplified model for the electrostatic field calculation in the contact area, see Fig. 13.1

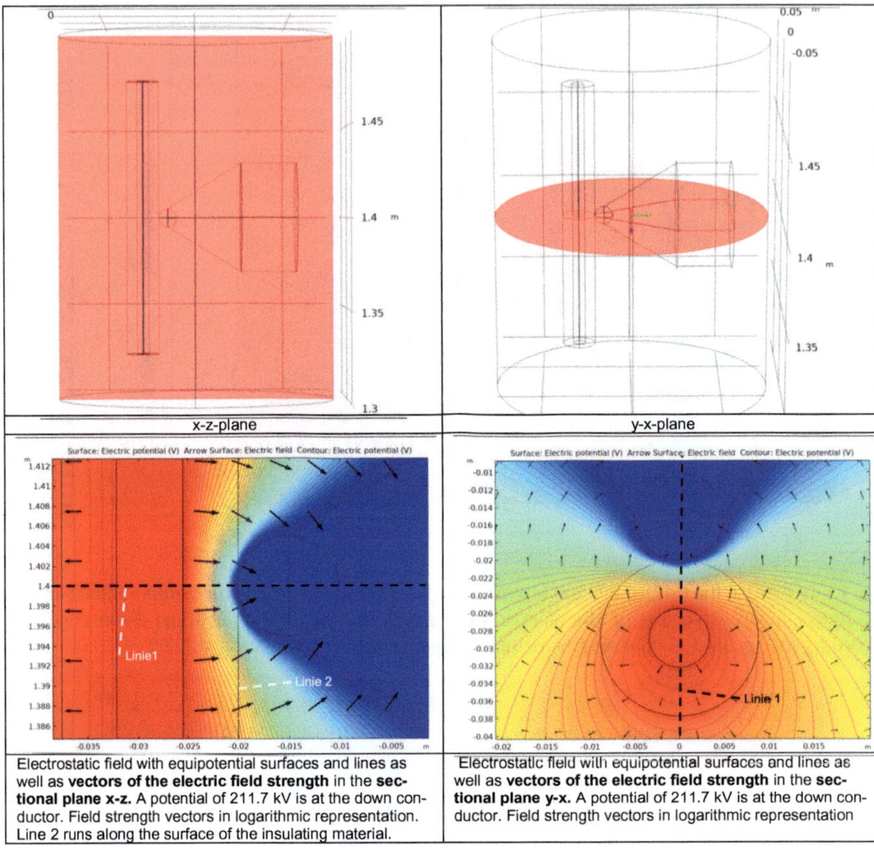

Fig. 13.6 Sectional planes and field plot of the arrangement according to Fig. 13.5

streamer formation is possible. However, a breakdown of the air gap to the person is unlikely because the down conductor is insulated and does not break through, see also Sect. 17.2.2 (Fig. 13.10).

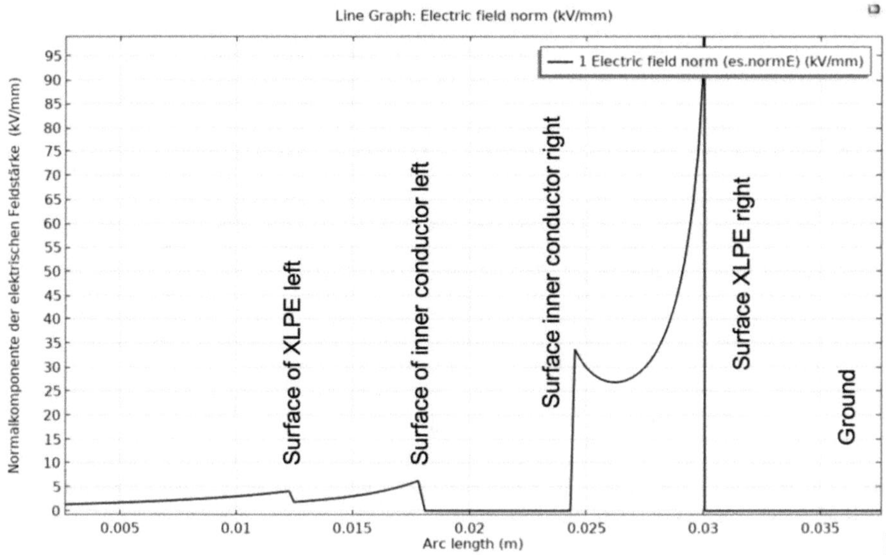

Fig. 13.7 Magnitude of electric field strength on the line 1 shown in Fig. 13.6

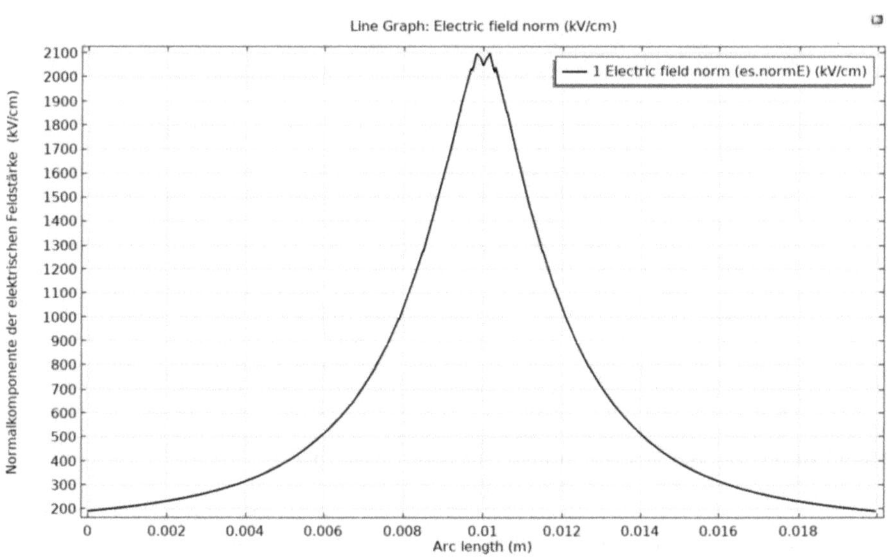

Fig. 13.8 Magnitude of the electric field strength on the line 2 shown in Fig. 13.**6**

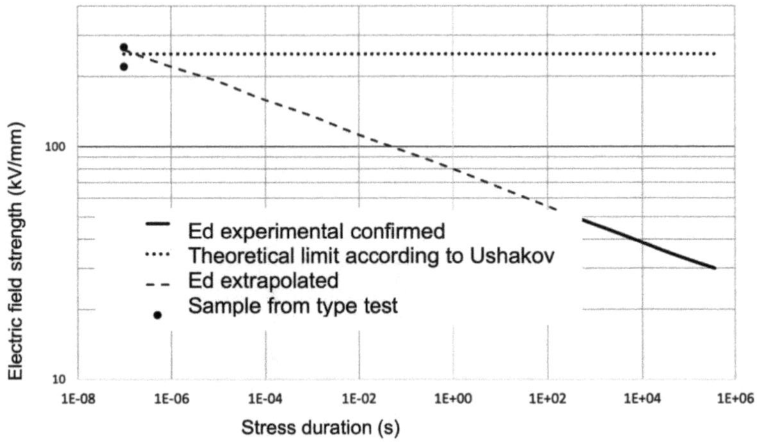

Fig. 13.9 Limits of electrical strength (in kV/mm) of XLPE (cross-linked polyethylene) according to Ushakov [60] and results from type tests [17]

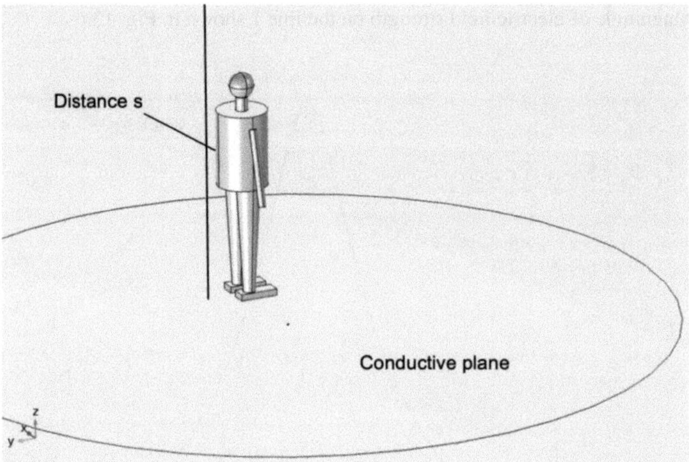

Fig. 13.10 Person with approach to an isolating down conductor

Persons approaching an insulated down conductor without touching it are therefore not exposed to any danger (Table 13.2).

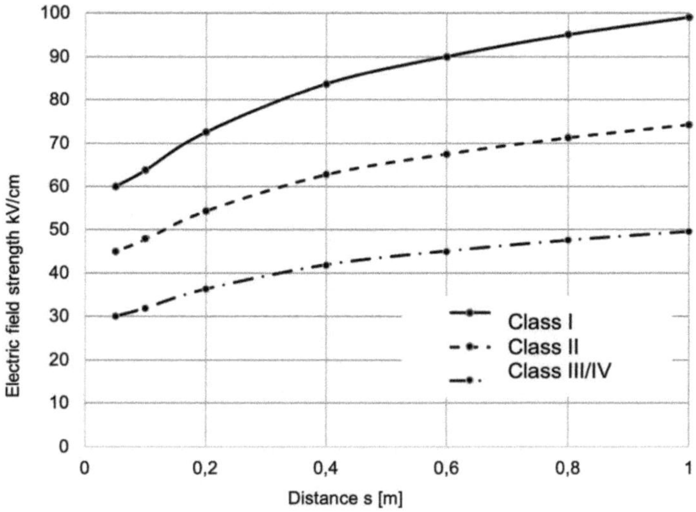

Fig. 13.11 Electric field strength on the surface of the insulating down conductor at a height of 1.4 m as a function of the distance s of the person

13.3 Type Test of Insulating Down Conductors

According to the state of the art, no type test has yet been specified for insulated down conductors **for personal protection.** Type tests can be carried out on the basis of the Technical Specification TS IEC 62561–8 standard until a valid test standard IEC 62561-9 is published. Such tested products are already on the market. Tests were also carried out also under rain conditions.

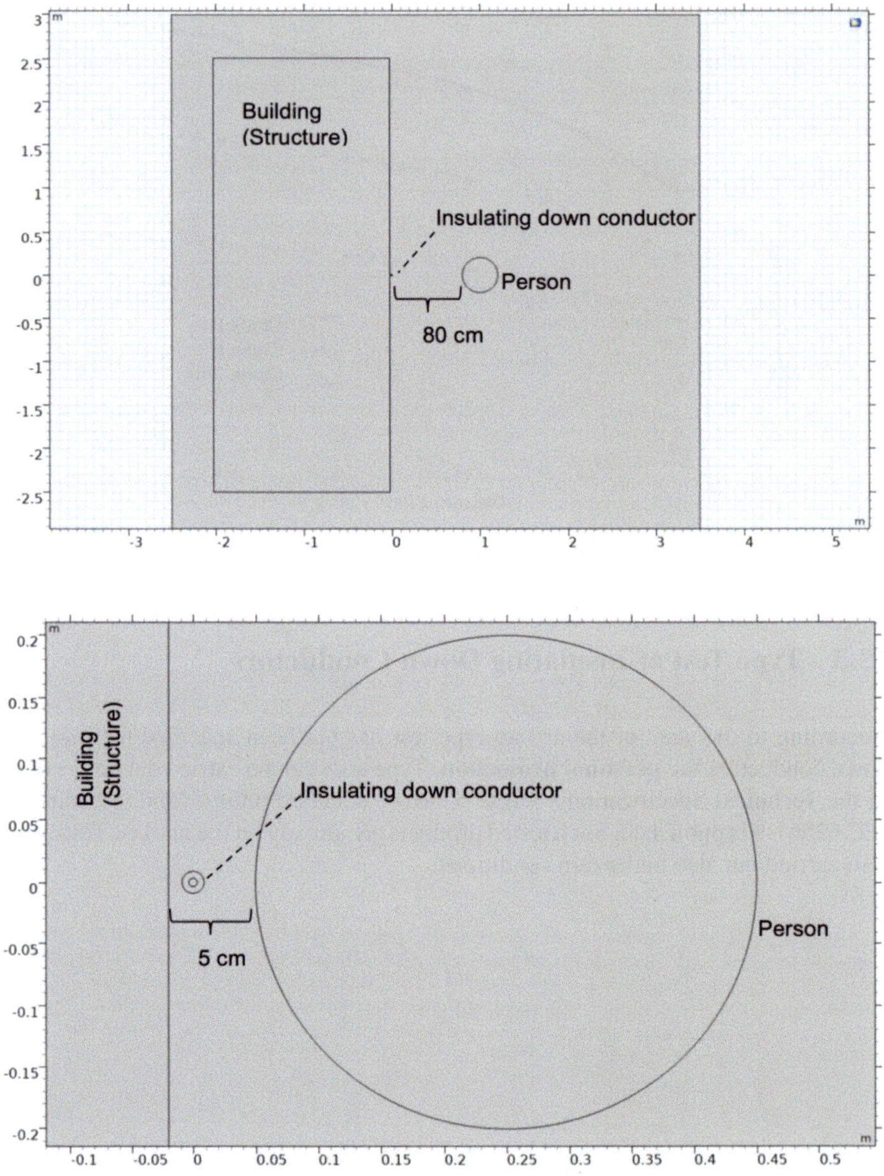

Fig. 13.12 2-D arrangement of an isolating down conductor in front of a building. (*Note* different scales)

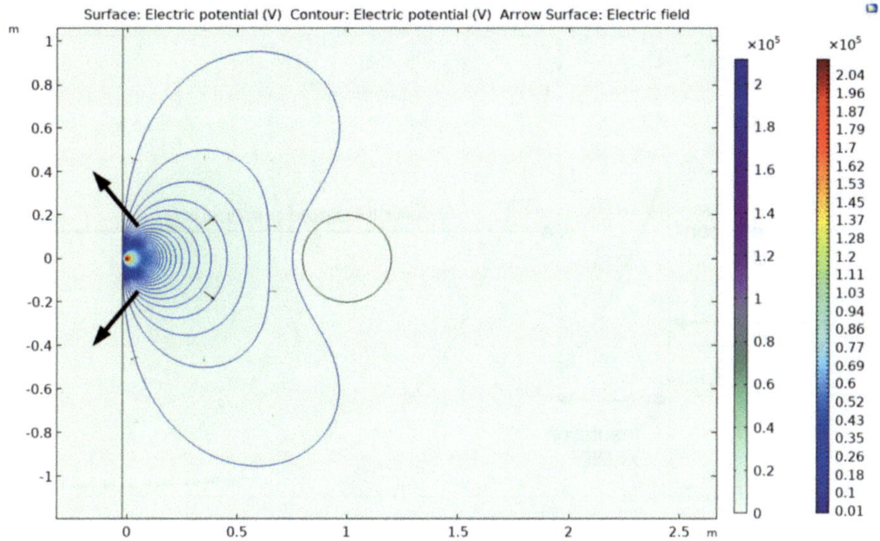

s = 0,8 m and 211,7 kV applied voltage on the insulating down conductor

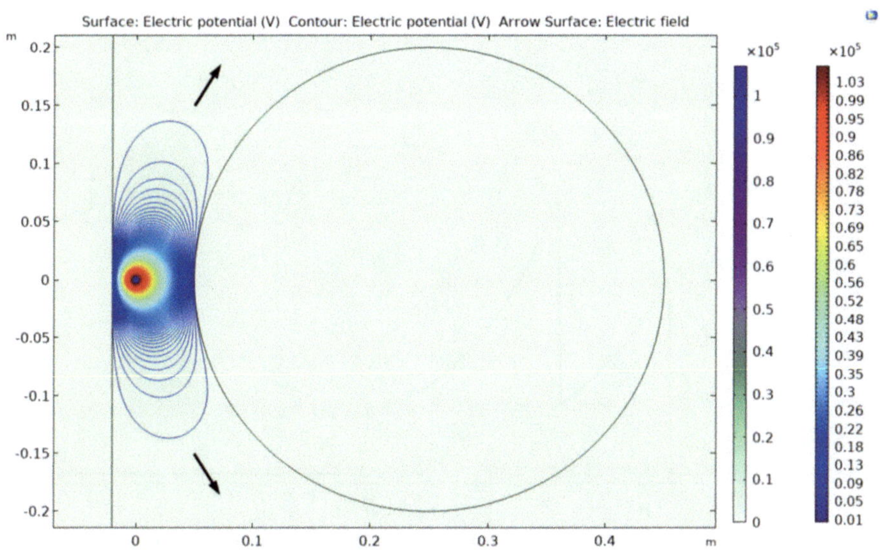

s = 5 cm and 107 kV applied voltage on the insulating down conductor

Fig. 13.13 Field with equipotential lines and areas with vectors of electric field strength

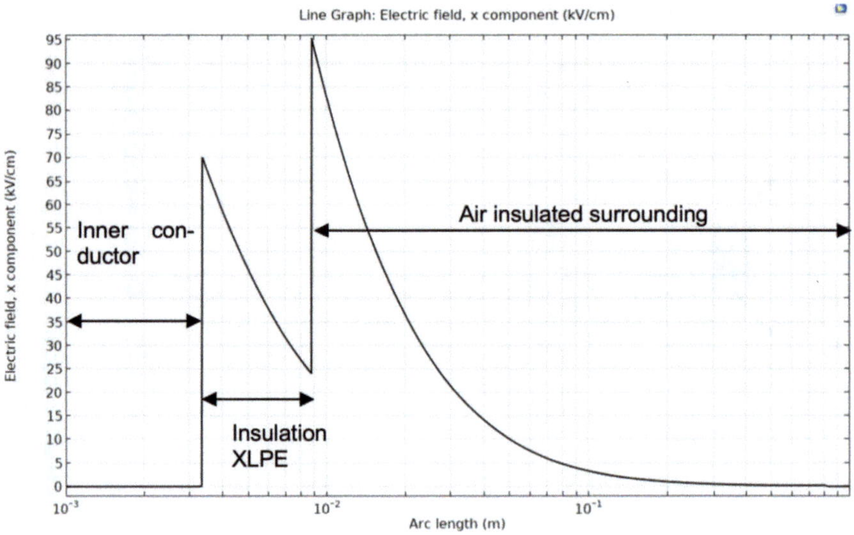

s = 0,8 m and 211,7 kV applied voltage on the insulating down conductor

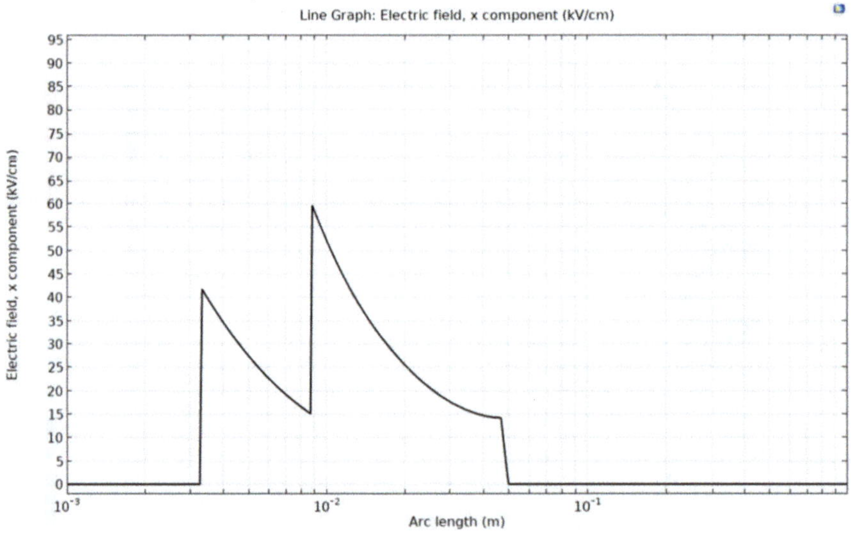

s = 5 cm and 107 kV applied voltage on the insulating down conductor

Fig. 13.14 Amount of the normal component of the electric field strength on the line between the down conductor and the person for two distances s

Table 13.2 Determined values for the numerical field calculation

Distance s	Mutual inductance M	Induced voltage M di/dt With di/dt according to IEC 62305–1, Table B.1 Values as per Table 13.1 for 0,25/100			Maximum field strength at the surface of the insulated down conductor at a distance s		
cm	μH	kV			kV/cm		
		I	II	III/IV	I	II	III/IV
100	1,596	220,25	165,18	110,25	99	74,25	49,5
80	1,534	211,70	158,78	105,85	95	71,25	47,50
60	1,454	200,69	150,51	100,34	90	67,50	45,00
40	1,341	185,06	138,80	92,53	83,7	62,78	41,85
20	1,149	158,57	118,93	79,28	72,5	54,38	36,25
10	0,960	132,49	99,36	66,24	63,8	47,85	31,90
5	0,775	106,95	80,22	53,48	60	45,00	30,00

This voltage also acts when a person touches the down conductor

Chapter 14
National and International Statistics of Deaths and Injuries

The available information from international sources [43–46] on the number of deaths and injuries is shown in Figs. 14.1 and 14.2. It is noticeable that the figures for injuries due to touch voltage are only in the 5% range. However, the figures do not allow a detailed indication of injuries/deaths at a lightning conductor. The figures for touch voltage in the sources also refer to cases in which a metal structure is hit and the resulting potential is transmitted e.g. through water pipes or wired devices (telephone). It can therefore be assumed that the number of injuries/deaths caused by direct contact with a down-conductor is very low. This can probably also be explained from a psychological point of view by the fact that a person who is close to a building in the thunderstorm field will enter it. If entry is not possible, it is unlikely that a person will stand near a down conductor and touch the down conductor with bare hands. In addition, it is extremely unlikely that a person touches the down conductor line with an outstretched hand. In the case where a person does not touch the arrester directly and stands only a few cm in front of the arrester, the induced voltage is also smaller because of the smaller area covered and the coupled energy and charge in the case of a flashover from the arrester to the person are correspondingly smaller.

Comparable information on lightning accidents in Germany can be found in the database on the website of the Federal Health Reporting Office, www.gbe-bund.de and is shown in Fig. 14.3. For a keyword search, enter e.g. lightning strike or the ICD-10 code X33. The ICD-10 classification has been in effect since 1998 until today. Until 1997, coding was done according to ICD-9. For the data from 1980 onwards, the ICD-9 code E907 is to be given as the keyword search. This shows a strong decrease in the number of persons killed by lightning since the 1980s. This trend has also been proven in international statistics since 1900.

According to [62], 2222 men and 929 women were killed by lightning in France between 1854 and 1884. This corresponds to an average number of 105 deaths per year. The number of injured per year is 5 times higher than the number of deaths per year. The high number is due to people working in the fields without protection and knowledge of the dangers of thunderstorms.

The VDE/ABB publishes statistics according to Fig. 14.4.

J. Meppelink, *The Lightning Rod as a Danger*,
https://doi.org/10.1007/978-3-031-30434-7_14

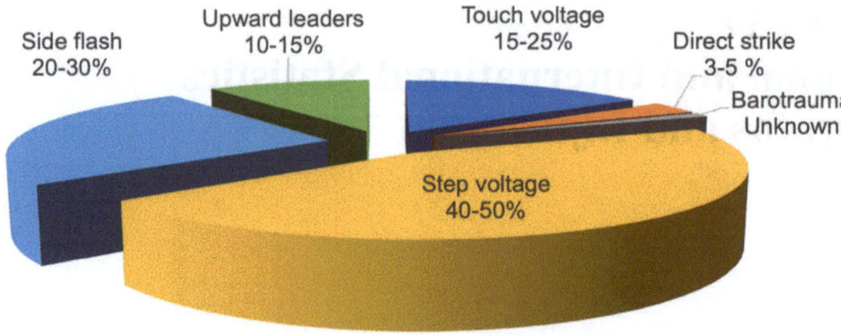

Fig. 14.1 Injury due to lightning [46]

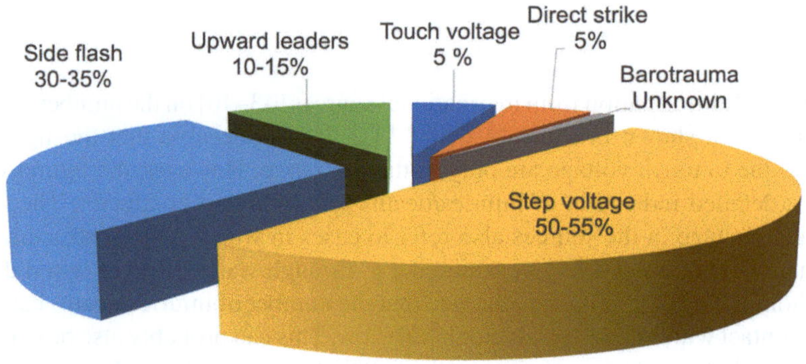

Fig. 14.2 Death due to lightning [45]

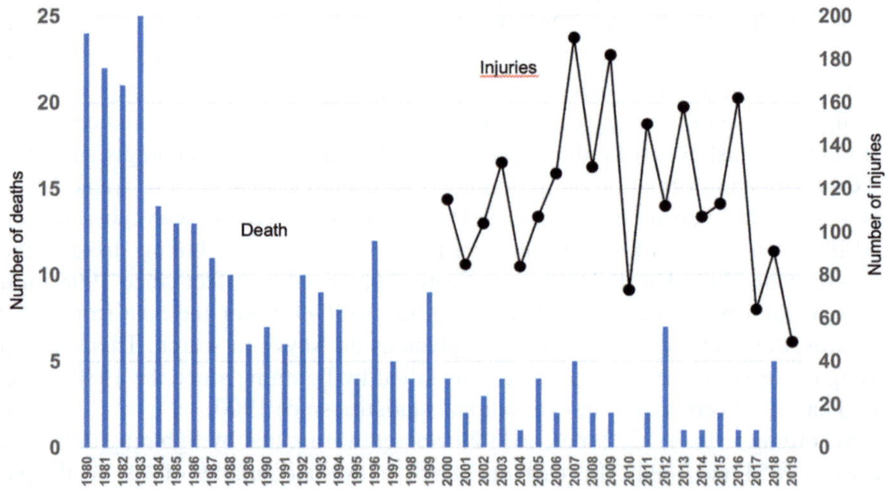

Fig. 14.3 Deaths and injuries as a result of lightning in Germany on the basis of www.gbe-bund.de

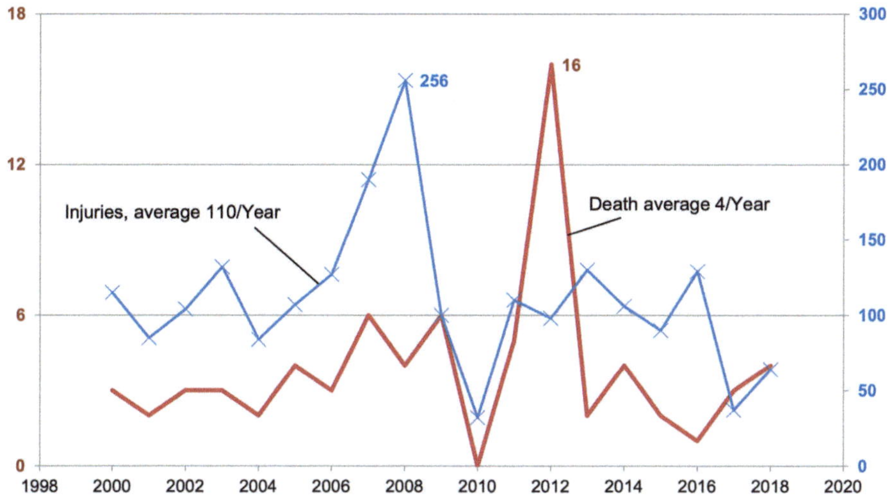

Fig. 14.4 Deaths and injuries as a result of lightning in Germany. *Source* VDE/ABB Unterausschuss Statistik

In the available statistics, it is not possible to extract the number of events caused by touching a down conductor. From the reports of the installers, however, no cases are known in which persons were endangered or died by touching the down conductor during a lightning strike.

As a conclusion, it can be stated that.

(a) Touching a person at the moment of a lightning strike is very unlikely.
(b) Other type of lightning impact pose a much higher risk.

If the risk of injury/death is to be completely eliminated, equipotential bonding, site insulation and the installation of high voltage insulated down conductors are a safe solution.

Another solution is the less practical restriction of access and the posting of warning notices. This is especially used for high-voltage pylons located on rocky ground.

Chapter 15
Statistics of Relevant Parameters of Lightning

The flash parameters are specified in detail in the standard [4]. In the following, only some aspects for the negative subsequent stroke are examined.

15.1 Statistics of the Steepness of Current of Negative Cloud to Ground Flashes

The basis for the calculation of the induced voltage is the standardised current slope of predominantly negative cloud-to-ground lightning according to [4]. Since the current slope is of great importance for a normative determination of the test of insulating down conductors for protection against contact, the origin of the data in [4] shall be examined below.

In [4]:

1. the current gradients actually measured by Berger are presented as a statistical distribution.
2. normative values for the maximum current slope within a lightning protection class (e.g.: 200 kA/μs for negative subsequent flashes and Class 1 and 100 kA/μs for negative first flashes and lightning protection Class 1).
3. current forms for the calculation (Heidler functions), whereby it is to be noted that with the Heidler functions a current steepness occurs that is 40% greater than the values specified in [4]. Furthermore, the current slope in the Heidler functions is proportional to the current peak values. This does not correspond to the measured values in [52]. Berger's measurements show that for negative sequence flashes, high current peak values are associated with low current slopes and, conversely, small current peak values are associated with high current slopes [52]. This is an important point in the overall hazard assessment.
4. no reference is made to the correlations of individual parameters with each other, but each parameter is considered on its own merits.

J. Meppelink, *The Lightning Rod as a Danger*,
https://doi.org/10.1007/978-3-031-30434-7_15

For further details, reference is made to a technical paper [52], from which some figures are reproduced below. The maximum current slope di/dt is shown in [52] as a function of the current peak value and shows the following behaviour.

1. the maximum current slope decreases with increasing current peak value !
2. the highest current slope was measured for negative sequence flashes at current peak values < 50 kA peak value.
3. the values end at a current slope of 100 kA/μs because the measuring equipment at the time of the measurements in the 1960s was not capable of measuring higher current slopes.
4. most of the registered flashes are negative sequence flashes (Fig. 15.1).

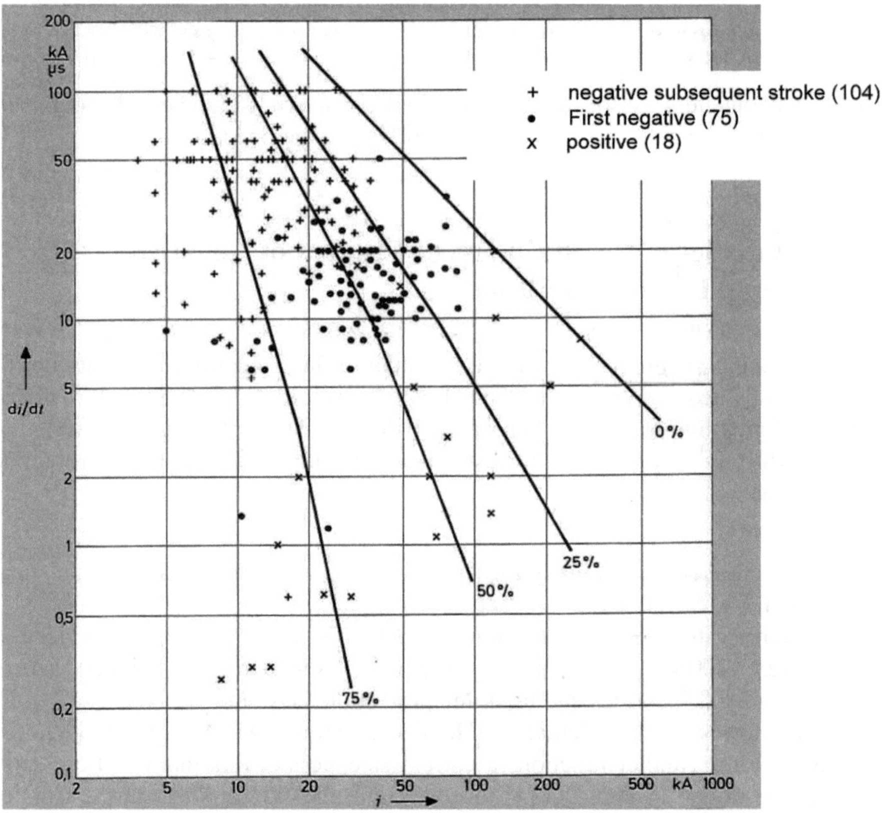

Fig. 15.1 Associated characteristic values of the three types of lightning current with regard to peak value and greatest slope di/dt in the current rise according to [52]. The values in brackets indicate the number of lightning events measured by Berger (added by the author)

15.2 Normative Values in IEC 62305–1

According to [52], due to the limited measurement technology in the 1960s, the maximum current slope for the lightning currents was set correspondingly higher, see Fig. 15.2 and defines the current lightning current parameters in [4]. It must be pointed out here that in [4] a constant current slope is simplistically assumed, which does not correspond to reality. This discrepancy was eliminated by introducing the analytical lightning current shapes. However, a new problem was created, since the analytical lightning current shapes exhibit a larger current slope than corresponds to the specified parameters for the current slope in [4].

In the informative annex of [4], analytical lightning current shapes (Heidler functions) are proposed for the purpose of analysis. In Figs. 15.3 and 15.4 the Heidler functions of the lightning currents for 0.25/100 and 1/200 with the respective differential quotients di/dt are shown.

In MT 8 (Maintenance Team 8 TC 81 deals with ED 3 of IEC 62305–1), the author already pointed out the discrepancy between the maximum di/dt in the Heidler function and the normative maximum current gradients and introduced a note in the future ED 4 of IEC 62305–1.

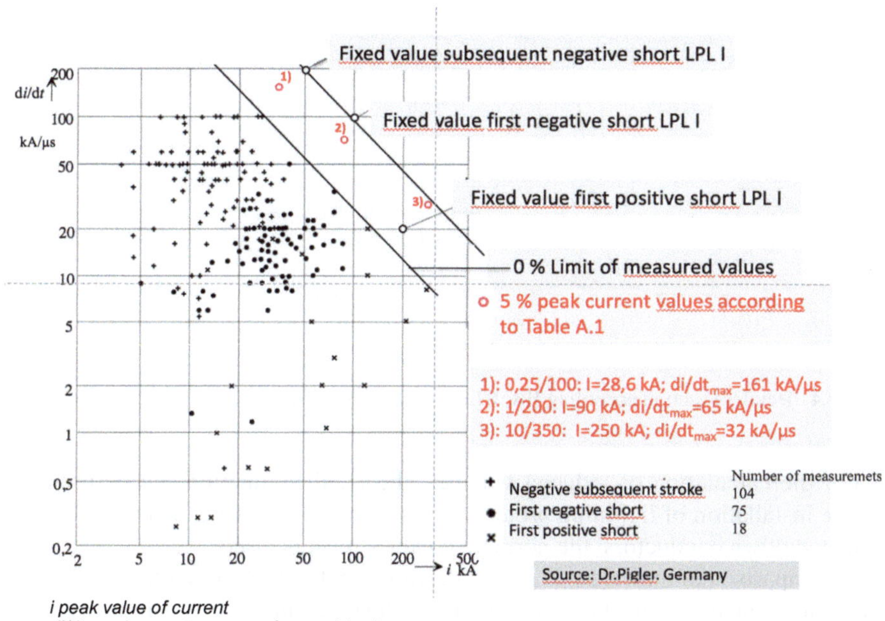

Fig. 15.2 Determination of the characteristic values of the three lightning current types according to IEC 62305–1 and measured values according to [52]. For comparison, the values for a 5% probability were added. The yellow marked field indicates the number of lightning events measured by Berger (added by the author)

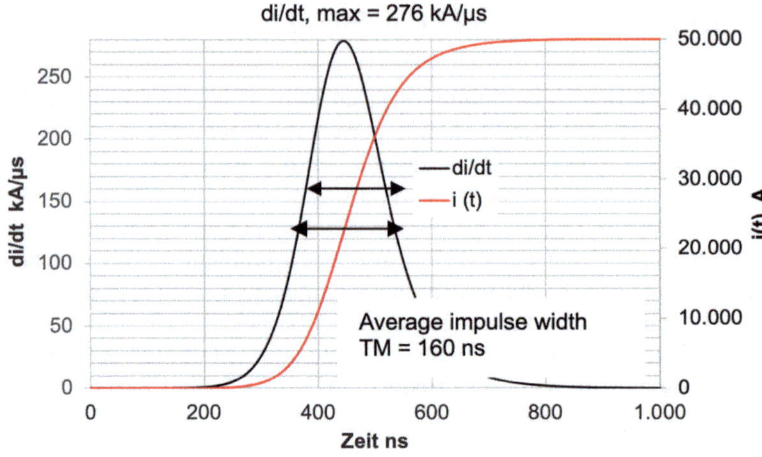

Fig. 15.3 Heidler function according to IEC 62305–1 for 0,25/100 Class I

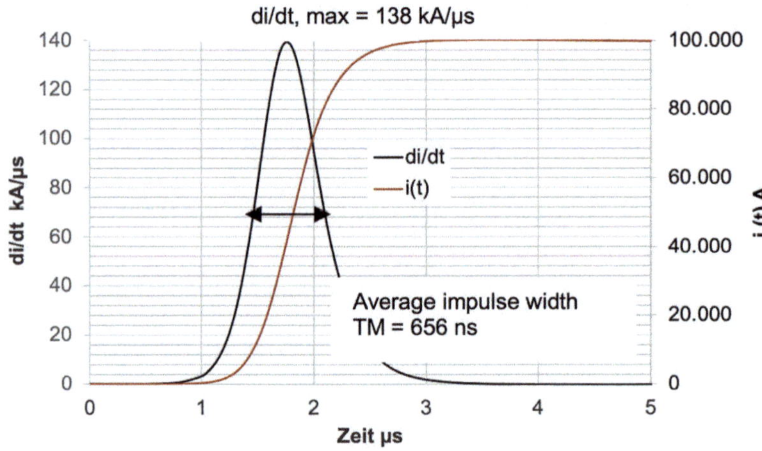

Fig. 15.4 Heidler function according IEC 62305–1 for 1/200 Class I

For the calculation of induced voltages, the Heidler function can be used, but for the installation of lightning protection systems and also when using type-tested insulated down conductors, the normative values must be used.

A comparison of the values is shown in Table 15.1, which shows that the Heidler functions (time functions) have 38% higher di/dt values and sine post-figures approx. 60% higher di/dt values.

Therefore, when discussing a test standard for insulated down conductors for touch protection, it is important to ensure that the values according to IEC 62305–1, A3 are used. Otherwise, an unrealistically high voltage is tested, which in reality never occurs at this level.

Table 15.1 Comparison of the di/dt values for the lightning currents according to IEC 62305–1, A3 and for comparison the Heidler functions and the values for a simulation by a sine function

	T_1	Peak value for LPLI	Average di/dt as per IEC 62305–1 A.3 for LPLI 1)	Max. di/dt as per time functions Annex B for LPLI	Max. di/dt as per sine wave for LPLI
	μs	kA	kA/μs	kA/μs	kA/μs
10/350 (first positive impulse)	10	200	20	27.3	31.4
1/200 (first negative impulse)	1	100	100	139.4	157
0,25/100 (subsequent negative impulse)	0.25	50	200	279	314

(1) These values are taken from the current divided by the front time T_1, assuming a linear rising impulse current as a first approximation.
Note The relations are the same for the other classes of LPL

In [4], a value of 200 kA/μs with a probability of 1% is specified for the current slope of the negative subsequent stroke in Class I, refer to Fig. A5 in IEC 62305–1, curve indicated with no. 15. For Class III/IV, 150 kA/μs, this results in a value of 2%. For Class III/IV a value of 100 kA/μs at a probability of 5% results. According to the scope of [4], these are also to be applied for the protection of persons.

Chapter 16
Calculation of Risk R_A for Death and Injury of Living Beings Due to Electric Shock as a Result of Touch- and Step Voltages According to IEC 62305–2

The following calculations according to the state of the art in IEC 62305–2 show a considerable discrepancy to the results of the calculations in this report.

For the building shown in Fig. 1.3, which was used as a basis for calculations in this report, a value of ND = 0.0029106 or a mean time between two lightning strikes of 34.35 years results according to the information in IEC 62305–2 for NG = 3 lightning strikes/(km2*year) and a CD value of 2 according to Table 16.1.

The probability of a negative subsequent lightning with a current gradient of 200 kA/μs is then about 1% for lightning protection class I and for 100 kA/μs about 5% for lightning protection class III/IV. This means that, on a static average, a negative subsequent lightning occurs every 3435 years with a current gradient of 200 kA/μs. In lightning protection class III/IV, a negative sequential lightning with a current gradient of 100 kA/μs then occurs every 687 years. It should also be taken into account that only about 50% of all negative downward flashes contain a sequential flash. Therefore, a negative downward flash with a current slope of 200 kA/μs would only occur every 6870 years and every 1374 years with 100 kA/μs.

Since a negative subsequent flash does not occur without a negative first flash, this must also be taken into account when considering the impact probability and it must also be noted that the energy and charge of both types of lightning have a cumulative effect with regard to ventricular fibrillation.

The probability of a negative first flash with a current of 100 kA is then about 3% for lightning protection class I and for 50 kA about 25% for lightning protection class III/IV. This means that on a static average, a negative first lightning with a current of 100 kA occurs every 1145 years. In lightning protection class III/IV, a negative first lightning with a current of 50 kA occurs every 137 years.

J. Meppelink, *The Lightning Rod as a Danger*,
https://doi.org/10.1007/978-3-031-30434-7_16

These figures already indicate that the probability of a hazardous situation is very low, even in lightning protection class III/IV. An estimate of the risk RA is given in Table 16.2 and evaluated for other parameters in Table 16.3. Even in the simplest case of an environment with arable land without an external lightning protection system and without further measures, applying the information in IEC 62305–2 results in a mean time MG = 1/RA between two hazardous events of 13 million years. This figure can probably be understood by the fact that:

(a) Contact of a person with a bare down conductor is very unlikely.
(b) Being in the vicinity of a down conductor (step voltage hazard) is highly unlikely
(c) Persons are standing at the time of the lightning strike at the down-conductor that was struck closest by the lightning (corner strike).

International data on lightning accidents, see Sect. 14, show that the number of lightning accidents due to contact is only 3–5% of all cases, but the number of lightning accidents due to step voltage is 30–35%. Therefore, there is a large discrepancy between the values calculated according to IEC 62305–2, the physical effects calculated in this report and the internationally available data on lightning accidents (Tables 16.1 and 16.2 and 16.3).

Table 16.1 Calculation of N_D and the mean time between two dangerous events M_G. For details, please refer to IEC 62305–2

Lage	A_D Collection area	N_G (1) Ground flash density	N_D Number of dangerous events	M_G (2) Mean time between two dangerous events
	$A_D = L*W + 2*3*H*(L + W) + \pi*(3*H)^2$		$N_D = N_G*A_D*C_D*10^{-6}$	$1/N_D$
		1/(km2*year)	1/Year	Years
Structure on hill top C_D = 2	= 15*15 + 2*3*10*(15 + 15) + 3,14*(3*10)^2 = 4581 m^2	3	= 3*4581*2*10^{-6} = **0,0,029,106**	1/0,0,029,106 = **34,35**

The value of 3 is an average value for conditions in Germany
This value is not defined in IEC 62305–2 and was introduced by the author for better understanding

Table 16.2 Estimation of the risk R_A according to IEC 62305–2 Sect. 6.2 for death and injury of living beings due to electric shock as a result of touch and step voltages for an office building 15 m × 15 m × 10 m using the example for arable land without lightning protection system. For details, please refer to IEC 62305–2

$R_A =$	N_D *	P_A *	L_A
	See Table 16.1	Probability of injury	Loss
		$\mathbf{P_A = P_{TA} * P_B}$	$\mathbf{L_A = r_t * L_T * n_z/n_t * t_z/8760}$
		$\mathbf{P_{TA}}$: Depends on additional protection measures $\mathbf{P_B}$: Depends on protection measures to reduce physical damage	r_t: factor reducing the loss of human life depending on the type of soil or floor L_T: typical mean relative numbers of victims injured by electrical shock(D1) due to one dangerous event n_z: number of persons in the zone n_t: total number of persons in the structure t_z: time in hours per year for which the persons are present in the zone See IEC 62305–2
		P_{TA}: 1, No protection measures $P_B = 1$ Structure not protected by LPS	$r_t = 10^{-2}$ agricultural or concrete $L_T = 10^{-2}$ All types $n_z = 10$ Person at entrance of structure $n_t = 100$ Person inside structure $t_z = 2920$ People working 8 h a day
$R_A =$	0,029,106/year	*1	* 10^{-2} * 10^{-2} * 10/100 * 2190/8760

$R_A = 7{,}27 * 10^{-8}$/year
$M_G = 1/R_A = 13.742.870$ years = Mean time between two dangerous events

$\mathbf{R_A}$: Component that relates to the injury of living beings. It is caused by electric shock as a result of contact and step voltages inside the structure and in an area up to 3 m around down conductors outside the structure. Type of damage L1 can occur and, in the case of agricultural installations, also type of damage L4 with possible loss of animals. *Source* IEC 62305–2

Table 16.3 Estimation of the mean time M_G between two events with injury and death of persons due to electric shock as a result of touch and step voltages for an office building 15 m × 15 m × 10 m for different variants. For details, please refer to IEC 62305–2

		Agricultural soil	Gravel	Asphalt
r_t		10^{-2}	10^{-4}	10^{-5}
		No LPS $P_{TA} = 1$; $p_B = 1$		
M_G	Mio years	13	137	13.742
		LPS Class III/IV $P_{TA} = 1$; $p_B = 0{,}1$		
M_G	Mio years	137	13.742	137.428
		LPS Class III/IV $P_{TA} = 0{,}1$ (Warning); $p_B = 0{,}1$		
M_G	Mio years	1.374	137.429	1.374.287
		LPS Class III/IV $P_{TA} = 0{,}01$ (3 mm XLPE); $p_B = 0{,}1$		
M_G	Mio years	13.742	1.374.287	13.742.871
		LPS Class III/IV $P_{TA} = 0$ (access restriction or reinforcement steel used as natural down conductor; $p_B = 0{,}1$		
M_G	Mio years	Infinite, zero risk		

r_t Factor, reducing the loss of human life depending on the type of soil or floor
P_{TA} Depends on additional protection measures
P_B Depends on protection measures to reduce physical damage
$M_G = 1/R_A$ Mean time between two dangerous events

Chapter 17
Strength of Air Gaps at Inductive Coupled Surge Voltages

To assess the danger when a person approaches the down conductor, it is necessary to measure or calculate the breakdown voltage of the insulating distance d with negative polarity according to Fig. 17.1. The applied voltage has a special shape according to Fig. 18.3, Sect. 19. This voltage shape does not correspond to the usual shape of the standard lightning impulse voltage 1.2/50 [56]. For the determination of the breakdown voltage with this special voltage form, few measurements are known from the literature [36, 47–51], which are examined in Sect. 17.1. This is followed by an overview of the calculation possibilities in Sect. 17.3.

17.1 Experimental Results from Literature

For the time range of interest here, Wiesinger [48] has investigated rod-plane arrangements in the range from 3 to 45 cm with negative polarity. These data are taken into account when calculating the safety distances (Figs. 17.2 and 17.3).

17.2 Calculation of Breakdown Voltage for Ramped Current Rise According to IEC 62305–1

17.2.1 Calculation Methodology

Die The strength of air spark gaps at steep surge voltages can be determined experimentally. From the results, laws can be derived from which an approximate calculation of the expected breakdown voltage of an arrangement can be expected even at steeper surge voltages. The calculation is based on Kind's voltage time area law and the methods derived from it [36, 47–51].

J. Meppelink, *The Lightning Rod as a Danger*,
https://doi.org/10.1007/978-3-031-30434-7_17

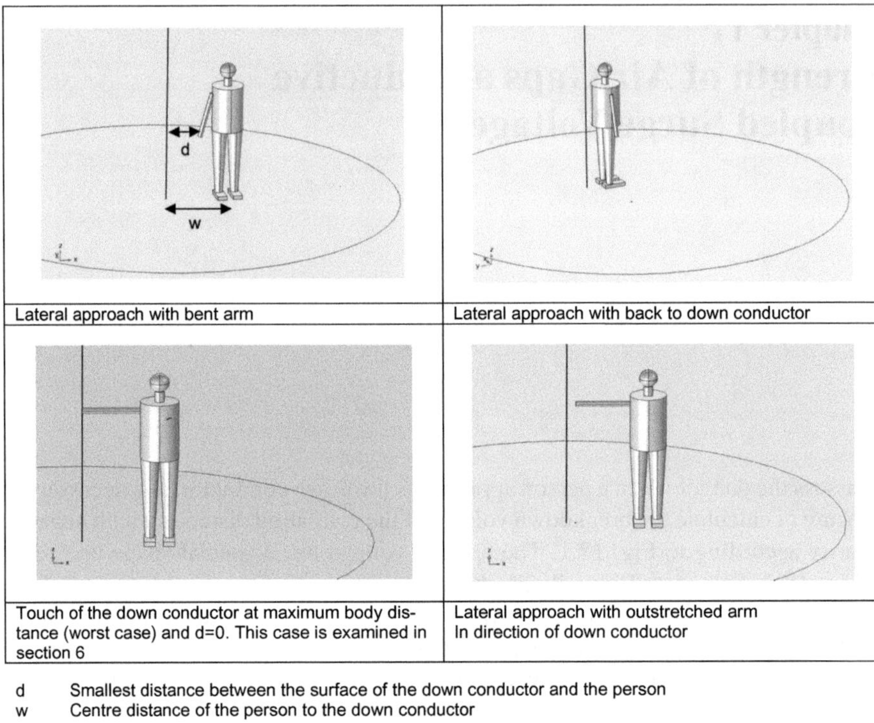

Fig. 17.1 Possible geometries in the approach of a person to the derivative

According to [36, 59], the breakdown voltage U_d can be calculated for a rod-plate arrangement according to Eq. 17.1. A linear current increase is assumed, Fig. 17.4, which then results in a rectangular voltage shape for inductive coupling. By applying the area-time law to a linearly increasing voltage, Eq. 17.1 is obtained for this case. Figure 17.5 shows measurements of the surge characteristics for three different electrode arrangements with a voltage form of 1.2/50 and a calculated surge characteristic according to Eq. 17.1.

$$U_d = \frac{d}{[m]} * 400kV * \left[1 + \frac{1}{T_1[\mu s]}\right] \qquad (17.1)$$

d Gap distance as per **Figure**, for rod plane geometry

T_1 Front time of the impulse voltage.

U_d Breakdown voltage.

An example shows for a gap distance of 25.5 cm a breakdown voltage according to Eq. 17.1 of

$$U_d = \frac{0,255}{[m]} * 400kV * \left[1 + \frac{1}{0,25}\right] = 510kV$$

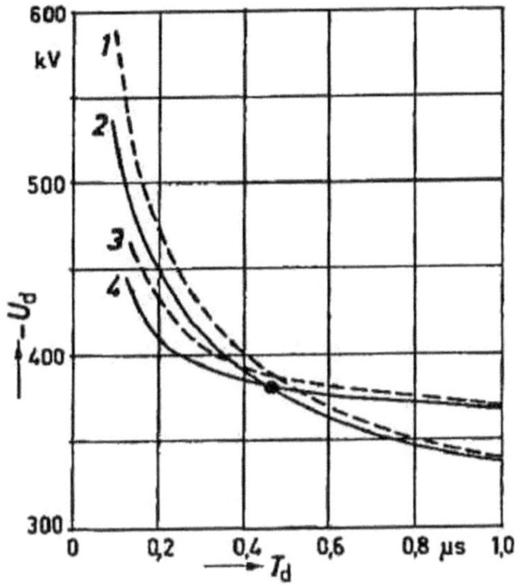

Ud	Average value of the negative breakdown voltage
Td	Time to breakdown
1	Plane-Plane gap a = 13,5 cm; d= 45 cm (corrected with air density)
2	As 1, but without correction
3	Rod-plane gap, plane on ground, a = 15 cm, (corrected with air density)
4	As 3, but without correction

Fig. 17.2 Voltage–time curves for steep front impulse voltage. *Source* [48]

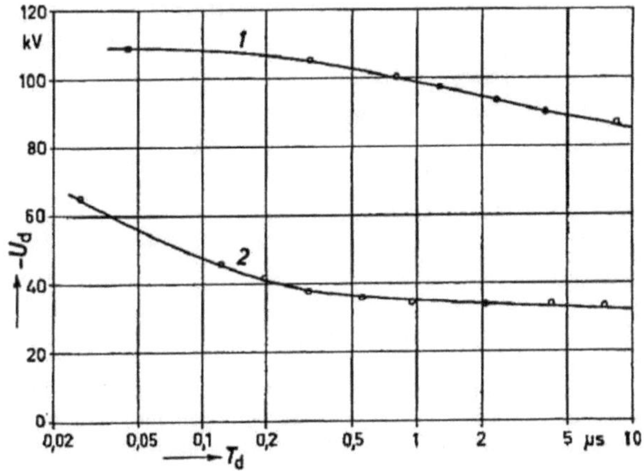

Ud	Average value of the negative breakdown voltage
Td	Time to breakdown
1	Rod-Plane gap, plane on ground a = 3 cm; (corrected with air density)
2	Surge arrester Category 10

Fig. 17.3 Voltage–time curves for steep front impulse voltage. *Source* [48]

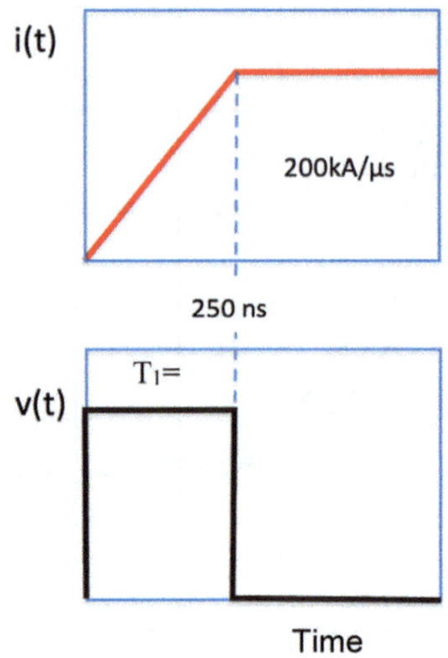

Fig. 17.4 Simplified approach to calculate the breakdown voltage for a ramp-shaped current rise with $T_1 = 250$ ns for negative subsequent stroke

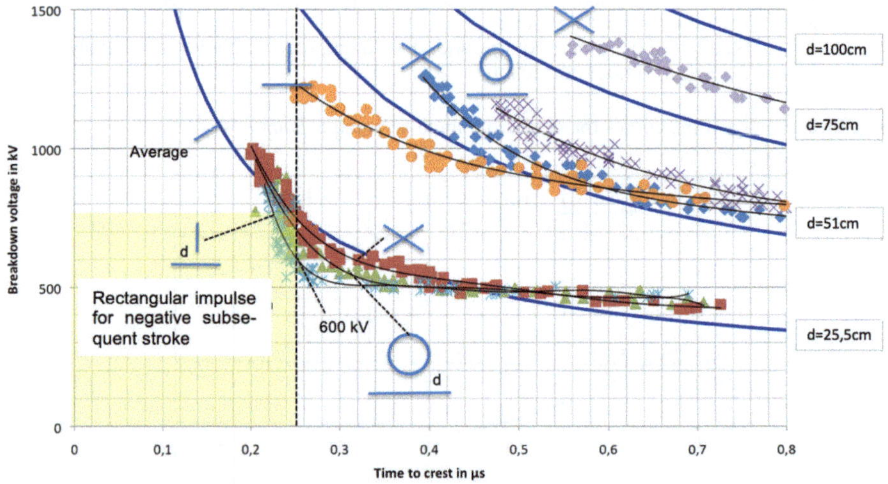

Fig. 17.5 Comparison between the impact characteristics calculated according to Eq. 17.1 and those measured for the stress form 1.2/50 of different arrangements with the impact distance d. Measurements were carried out by the author at the TU-Darmstadt

The possibility of measuring an experimental surge characteristic with a linearly increasing voltage is only possible with great effort. Therefore, a standardised lightning impulse voltage of 1.2/50, which is usually available in every laboratory, is used.

The breakdown voltages measured in the laboratory of the TU-Darmstadt in Fig. 17.5 deviate more or less from the theoretical curve according to Eq. 17.1. Furthermore, it can be seen that the deviations between measured and calculated impact characteristics increase with increasing impact distance.

17.2.2 Calculation of the Dangerous Zone for Induced Voltage with Ramp-Shaped Current Rise According to IEC 62305–1

With this knowledge, the danger zone in which a person should not stand next to a down conductor can be calculated. The induced voltage is given by M*di/dt. However, it should be noted that the length-related mutual inductance M' according to Eq. 17.2 increases with the logarithm of the distance ratio as the distance of the person from the down-conductor increases. The length-related value of the mutual inductance M' is shown in Fig. 17.7 for the usual dimensions of arresters. In general, the internal inductance of the down-conductor can be neglected because of the current displacement in the down-conductor. For the usual conductor diameters of down conductors in lightning protection systems, the conductor diameter does not play a significant role. Only when assuming that a lightning current flows through a downpipe does the large diameter result in a significantly smaller value for M' and thus also a smaller induced voltage (Fig. 17.6).

$$M' = 0,2 * \ln((R1 + b)/R1) \text{ in } \mu H/m \tag{17.2}$$

Die The breakdown voltage of the arrangement with the striking distance d in Fig. 17.2 is determined according to Eq. 17.1 and the induced voltage Ui is determined with Eq. 17.2 with a length of the down conductor of 1.4 m with M = M'*1.4 m where M' depends on the distance of the person to the down conductor. Both functions are shown for the three lightning protection classes for determining the danger zone in Figs. 17.9, 17.10 and 17.11. A surprising result emerges (Fig. 17.8):

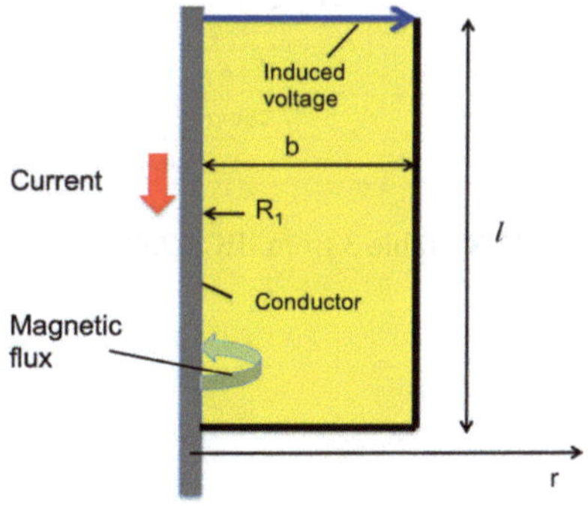

Fig. 17.6 Sketch for calculating the length-related mutual inductance M*

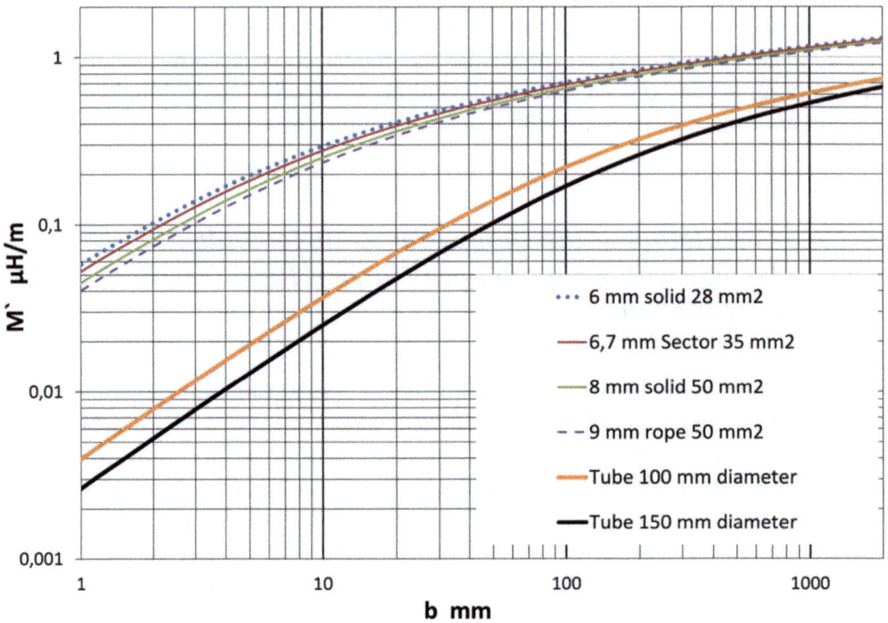

Fig. 17.7 Length-related mutual inductance of a conductor loop according to Fig. 17.6

Table 3 – Maximum values of lightning parameters according to LPLs

Current parameters		Symbol	Unit	LPL			
				I	II	III	IV
First positive short stroke	Peak current	I	kA	200	150	100	
	Charge	Q_{SHORT}	C	100	75	50	
	Specific energy	W/R	MJ/Ω	10	5,6	2,5	
	Time parameters	T_1/T_2	µs/µs	10 / 350			
First negative short stroke [a]	Peak current	I	kA	100	75	50	
	Average steepness	di/dt	kA/µs	100	75	50	
	Time parameters	T_1/T_2	µs/µs	1 / 200			
Subsequent short stroke	Peak current	I	kA	50	37,5	25	
	Average steepness	di/dt	kA/µs	200	150	100	
	Time parameters	T_1/T_2	µs/µs	0,25 / 100			
Long stroke	Charge	Q_{LONG}	C	200	150	100	
	Time parameter	T_{LONG}	s	0,5			
Flash	Charge	Q_{FLASH}	C	300	225	150	

[a] First negative stroke is only relevant for calculations and not testing.

Fig. 17.8 Table 3 from IEC 62305–1

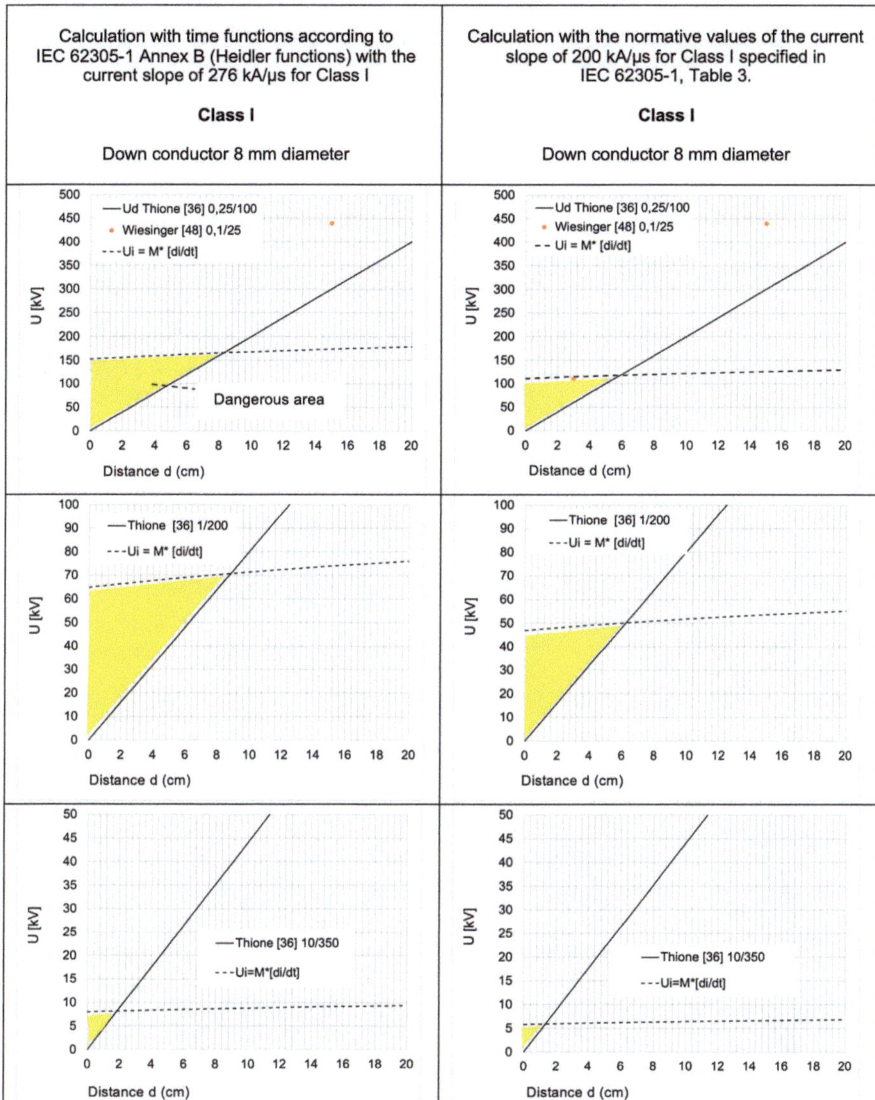

Fig. 17.9 Comparison of the danger areas for a person standing at a distance d from a down conductor. Lightning protection Class I and data from Table 3.1. *Note* Distance d means "distance of a person to the down conductor"

1. when using the time functions according to IEC 62305-1, Annex B (Heidler functions), there is practically only in lightning protection Class I for the lightning current 1/200 and a 9 cm hazard area. When using the normative values of the current slope in IEC 62305-1, **Table 3**, Fig. 17.8, the danger zone is reduced to only 6 cm.

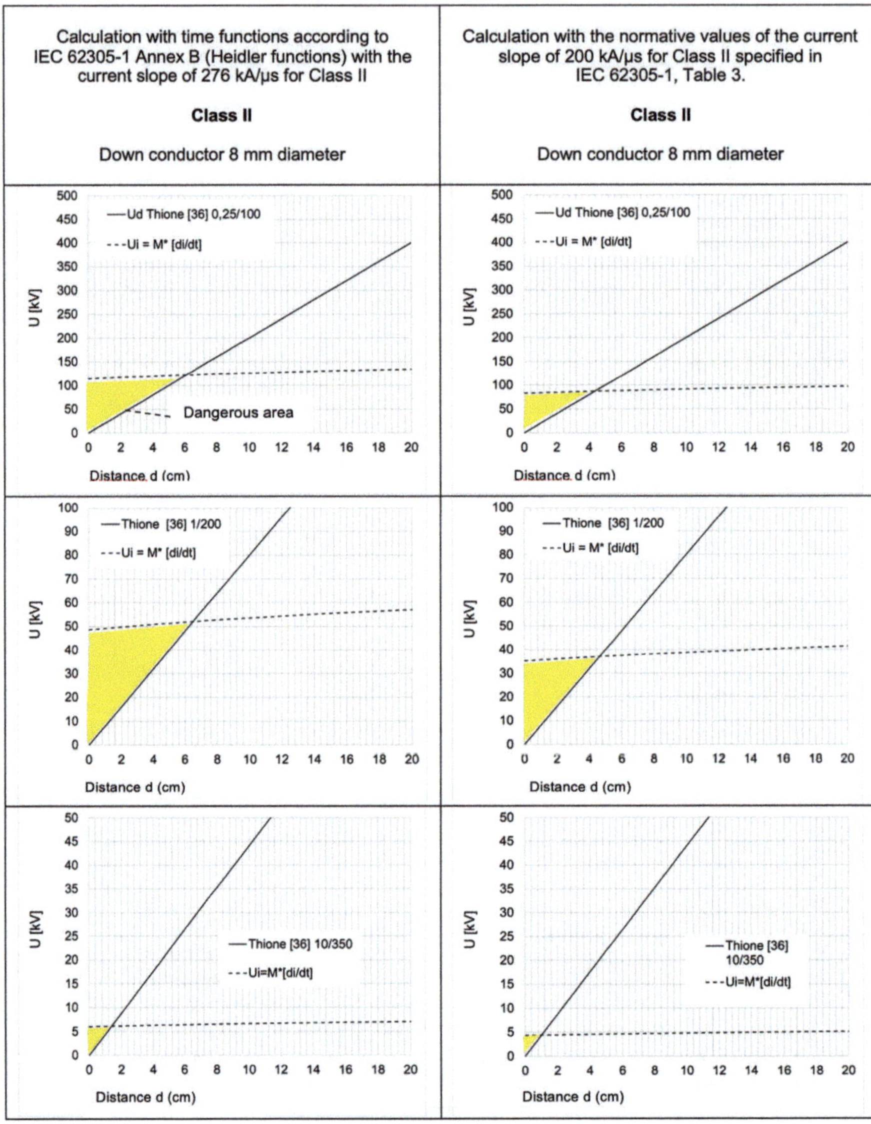

Fig. 17.10 Comparison of the danger areas for a person standing at a distance d from a down conductor. Lightning protection class II and data from Table 3.1. *Note* distance d means "distance of a person to the down conductor"

2. When using the time functions according to IEC 62305-1, Annex B (Heidler functions), there is practically only a 4 cm danger zone in lightning protection class III/IV for the lightning current 1/200. When using the normative values of the current slope in IEC 62305-1, **Table 3**, Fig. 17.8, the danger zone is reduced to only 3 cm.
3. If a person approaches a rainwater downpipe through which a lightning current is flowing, there is practically no longer any danger due to the counter-inductance M according to Fig. 17.7, which is only about half as large and thus half as large as the induced voltage.
4. The only practically relevant danger is therefore direct contact with the down conductor.
5. The measurements by Wiesinger [48] in Fig. 17.9 lie outside the calculated curves. The reason for this is the larger voltage slope used by Wiesinger.

17.3 Calculation of Breakdown Voltage for Induced Voltage (Delta Impulse)

The shape of the induced voltage is shown in Fig. 18.3. It is a so-called delta impulse with a pulse width of 160 ns only.

Pulses with a voltage level of several 100 kV to MV are applied in the military sector. Available generators are described in IEC TR 61,000–4-32:2002. These generators are normally not available for industrial test houses.

17.3.1 Test Facilities for EMP

Figure 17.12, 17.13 and 17.14 show two examples of EMP generators. The voltage rises rapidly, but drops back to zero very quickly. Similar generators are used to test the electromagnetic compatibility of complete aircraft.

17.3.2 Method of Calculation

The calculation of the breakdown voltage for the voltage form according to Fig. 19.3 using the area-time law is described in [53]. This results in a higher dielectric strength by a factor of 1.68 compared to a rectangular impulse, compare Sect. 17.2. This results in the impulse withstand voltage shown in Fig. 17.15.

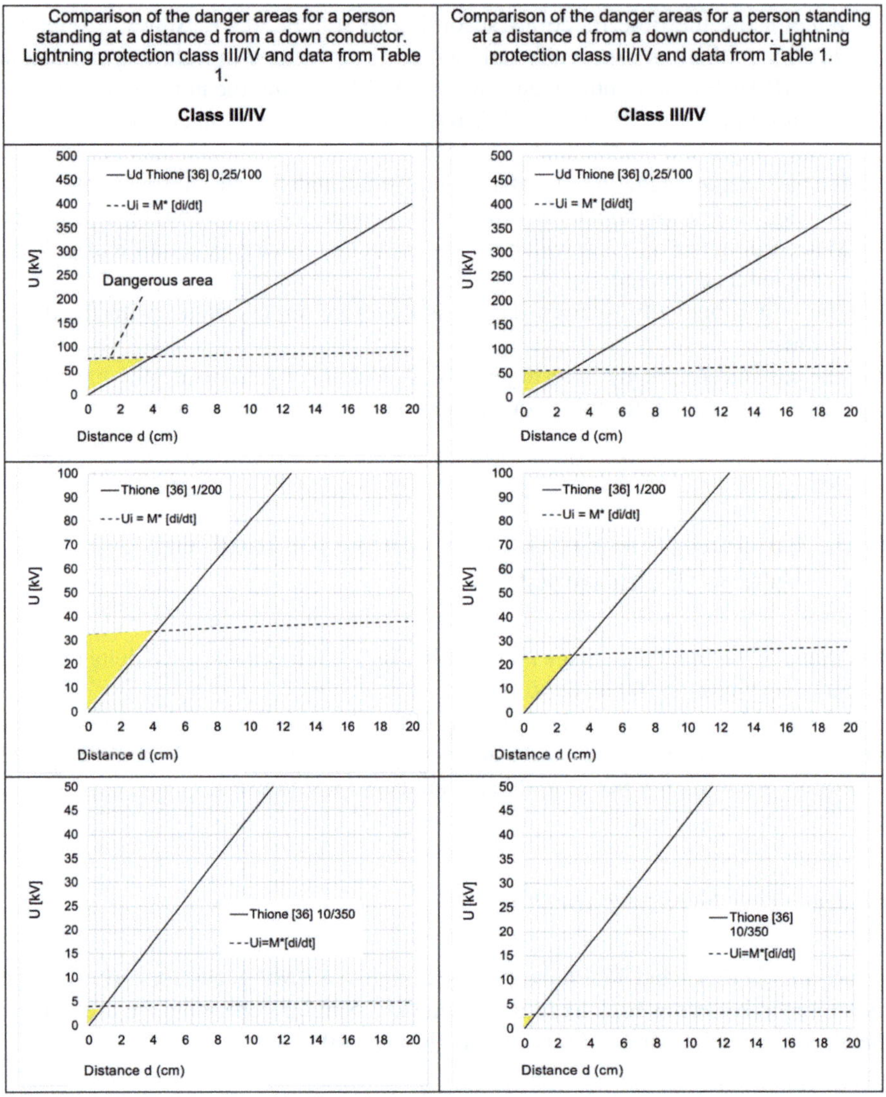

Fig. 17.11 Comparison of the danger areas for a person standing at a distance d from a down conductor. Lightning protection Class III/IV and data from Table 3.1. *Note* Distance d means "distance of a person to the down conductor"

17.3.3 Calculation of the Dangerous Zone for Induced Voltage (Delta Impulse)

As already explained in Sect. 17.2.2, the dangerous zone in Fig. 17.16 is evaluated using the same methodology, whereby the values from Fig. 17.15 apply as the impulse

Fig. 17.12 Pulse generator for 600 kV 5/50 ns at the University of Stuttgart. *Source* Dr. Köhler Universität Stuttgart

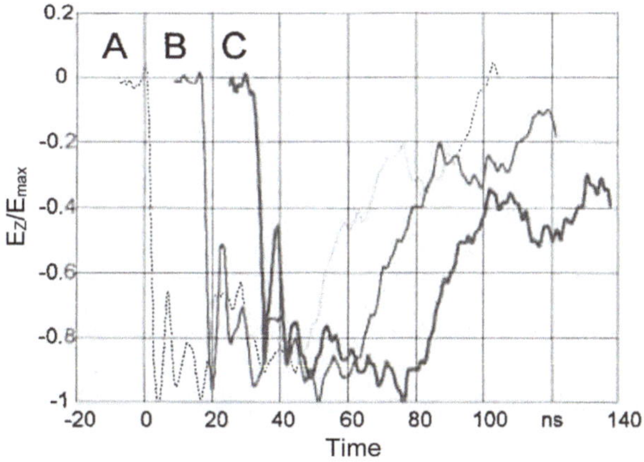

Fig. 17.13 Normalized E-field at 3 measuring points in 1 m height. Courtesy Universität Stuttgart, Professor S.Tenbohlen

withstand voltage. Therefore, even for lightning protection class I, the hazard area of 1.5 cm is significantly reduced compared to Fig. 17.9 due to the greater strength with delta pulse shape. It follows that for the other classes there is practically no more danger when approaching a down conductor, except for direct contact. The measurements of Wiesinger [48] also fit the calculated curves here, since the voltage slopes are in the comparable range.

Fig. 17.14 E-4 advanced airborne command post with EMP simulator. *Source* http://commons.wikimedia.org/wiki/Boeing_E-4?uselang=de

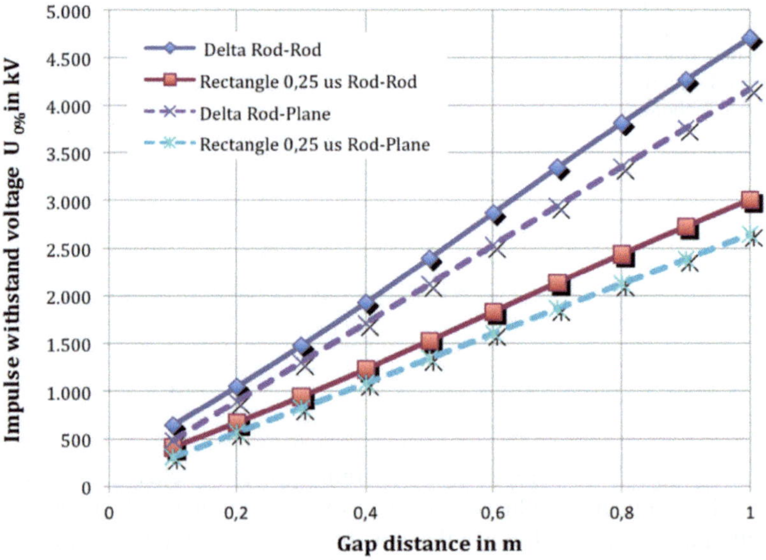

Fig. 17.15 Impulse withstand voltage of a rod-rod and a rod-plate arrangement for delta impulse according to Fig. 19.3. *Source* Meppelink ICLP Uppsala [53]

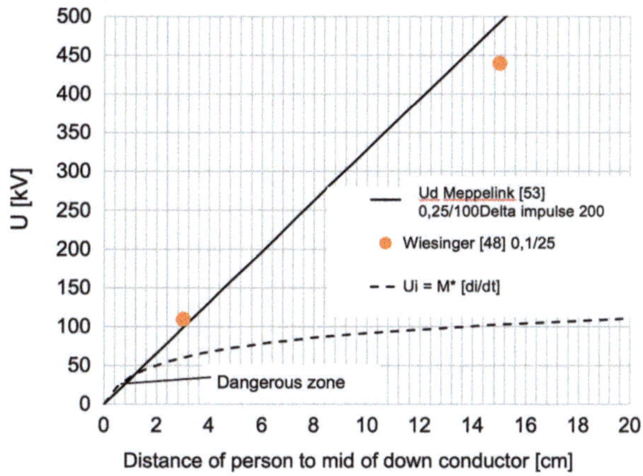

Fig. 17.16 Dangerous zone for a person standing at a distance d from the centre of a down conductor. Lightning protection class I and data from Table 3.1. Calculation for the values of the current slope of 200 kA/μs specified according to IEC 62305–1 and a U_d - calculation for the actual course of the voltage corresponding to the derivative of the negative sequence lightning stroke current according to [48]

Chapter 18
Numeric Calculation

18.1 Numeric Field Calculation Using Comsol-Multiphysics and XGS-Lab

The following table gives an overview of the possible calculations with the programmes used.

	Comsol-multiphysics	XGSA-TD and XGSA-FD
Time-domain	Electric currents, displacement current and eddy currents Currents Potentials Electric fields No magnetic (inductive) effects Any geometries and layers	Complete electromagnetic simulation with electric, magnetic fields, displacement current and magnetic coupling between conductors. Multilayer earth layers Currents Potentials Electric and magnetic fields Step and touch voltage along earth surface Limited calculation of thin layers, e.g. 2 cm water layer

(continued)

J. Meppelink, *The Lightning Rod as a Danger*,
https://doi.org/10.1007/978-3-031-30434-7_18

(continued)

	Comsol-multiphysics	XGSA-TD and XGSA-FD
Frequency-domain	Electromagnetic fields, electric currents, displacement current, eddy currents, inductive effects Currents Potentials Electric and magnetic fields Arbitrary geometries and layers	Complete electromagnetic simulation with electric, magnetic fields, displacement current and magnetic coupling between conductors Multilayer earth layers Currents Potentials Electric and magnetic fields Visualisation of dangerous areas for step and touch voltage along earth surface Limited calculation of thin layers, e.g. 2 cm water layer

Chapter 19
Applied Pulse Shapes 0.25/100 According to IEC 62305-1

The current pulse shape for the negative sequence flash defined according to IEC 62305-1 and its time derivative are listed below for information (Fig. 19.1).

$$i(t) = \frac{I}{k} \cdot \frac{\left(\frac{t}{\tau_1}\right)^{10}}{1 + \left(\frac{t}{\tau_1}\right)^{10}} \cdot e^{-\frac{t}{\tau_2}} \tag{19.1}$$

The time derivative of the pulse current i′ = di/dt according to Eq. 19.2 is required for the calculation of induced voltages and is shown as an example in Figs. 19.2, 19.3 and 19.4.

$$i' = \frac{I}{k} \cdot \left[\frac{10 \cdot \frac{t^9}{\tau_1^{10}}}{\left(1 + \left(\frac{t}{\tau_1}\right)^{10}\right)^2} \cdot e^{-\frac{t}{\tau_2}} + \frac{\left(\frac{t}{\tau_1}\right)^{10}}{1 + \left(\frac{t}{\tau_1}\right)^{10}} \cdot \left(-e^{-\frac{t}{\tau_2}}\right) \right] \tag{19.2}$$

J. Meppelink, *The Lightning Rod as a Danger*,
https://doi.org/10.1007/978-3-031-30434-7_19

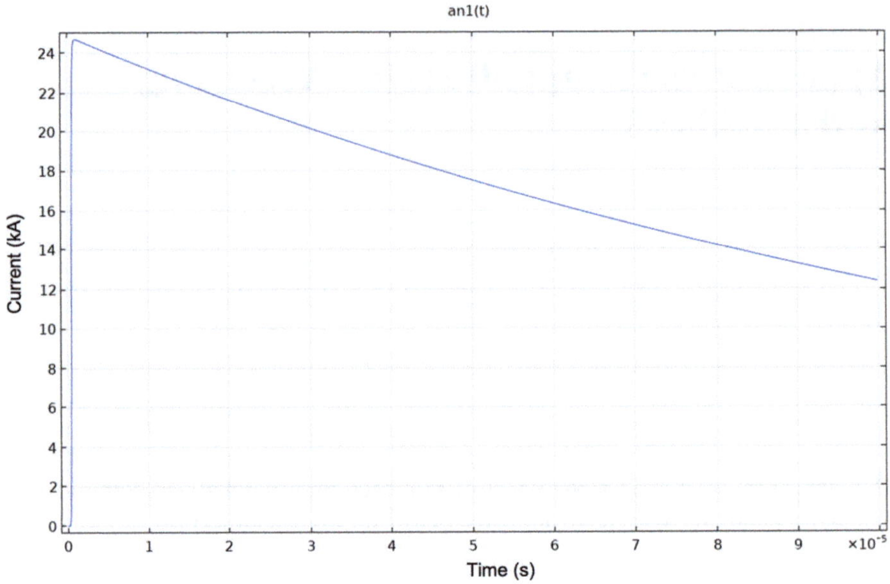

Fig. 19.1 Time characteristic of the current pulse 0.25/100 according to IEC 62305-1, for peak current values according to Table 3.1

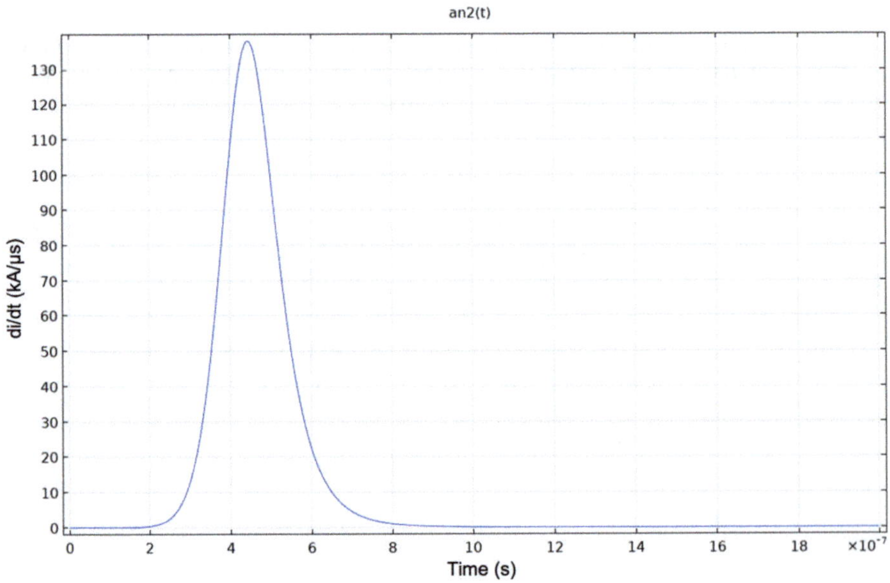

Fig. 19.2 Time characteristic of the differentiated current pulse 0.25/100 according to IEC 62305-1 for peak current values according to Table 3.1

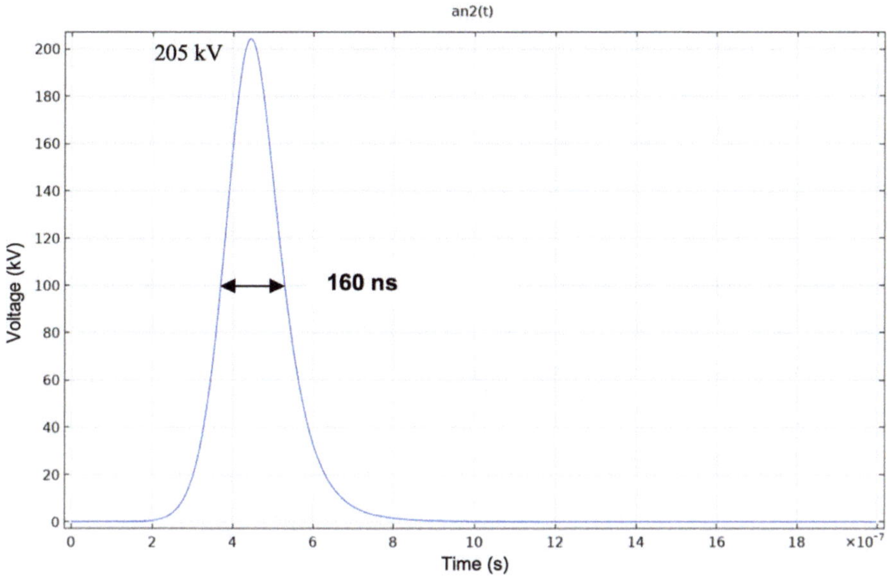

Fig. 19.3 Time curve of the induced voltage for M = 1.48 μH. Pulse shape 0.25/100 according to IEC 62305-1 for peak current values according to Table 3.1 for 8 mm diameter arrester, corresponding to 50 mm^2 cross-section

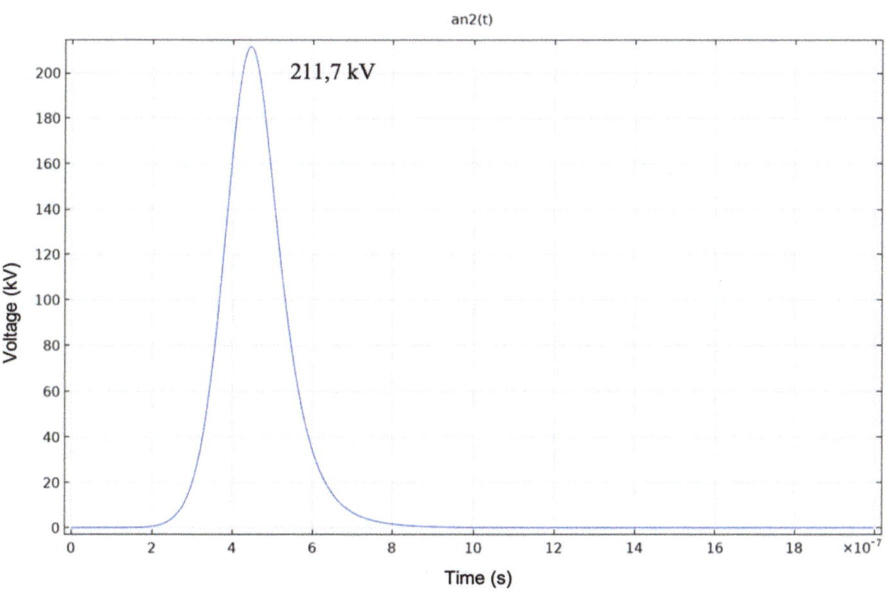

Fig. 19.4 Time curve of the induced voltage for M = 1.534 μH. Pulse shape 0.25/100 according to IEC 62305-1 for peak current values according to Table 3.1 for insulated down conductor with 6.7 mm diameter, corresponding to 35 mm^2 cross-section

Chapter 20
Propagation and Velocity of Surface Discharges

20.1 Surface Discharges in Nature

Surface discharges on insulated surfaces are known from the literature. An example of a discharge on a concrete slab is shown in Fig. 20.1, but no further details are available. A laboratory recording of a surface discharge is shown in Fig. 20.2, which occurred at a current of 20 kA with a rise time of 5–10 μs and falls within the range of the values investigated in this paper. It is therefore reasonable to assume that surface discharges occur at the foot of a person standing on asphalt.

20.2 Theory of Surface Discharges

The surface discharge was first described by M. Toepler. In this process, a single metal pole rests on an insulating plate with a metal coating on the back. A typical arrangement to explain the physical processes is shown in Fig. 20.3. The condition for the formation of surface discharges is a strongly pronounced component of the electric field strength E_N, which is perpendicular to the air/insulating material interface. The specific surface capacitance c_{OB} is decisive for the generation of surface discharges. If the applied voltage exceeds the initial voltage at the metal pole, pre-discharges are formed at the metal pole. This generates charge carriers that cannot flow off to earth because of the insulating material and are held on the surface by the electric field. This results in discharges, the so-called current threads on the surface. The insulator surface is charged from the metal pole with the polarity of the metal pole by successively igniting current filaments. In the process, the surface is covered with the surface charge Qob. The surface charge Qob forms a capacitor with the counter electrode, represented by the surface capacitance C_{OB}. The surface charge lies on a fictitious metallic surface. The specific surface capacitance c_{OB} is calculated from this fictitious metallic surface with the counter-electrode.

J. Meppelink, *The Lightning Rod as a Danger*,
https://doi.org/10.1007/978-3-031-30434-7_20

Fig. 20.1 Traces of a surface discharge on a concrete surface. *Source* Mr. Heuschen, VDE/ABB User Conference

Fig. 20.2 Traces of a surface discharge of about 4 m radius in the vicinity of a figure-centre earth rod on a sand surface in the laboratory with specific earth resistance of 270 Ohmm at an injected current of 20 kA with 5–10 μs rise time. *Source* Approved by Mr. Liew Ah Choy, EquiVolt M Pte Ltd. Singapore

$$c_{OB} = \frac{C_{OB}}{A} = \frac{\varepsilon_0 \varepsilon_r}{s} \tag{20.1}$$

The current in the discharge channel is given by

$$i_K = \frac{dQ_{OB}}{dt} = \frac{d[U * C_{OB}]}{dt} = C_{OB} * \frac{dU}{dt} + U * \frac{dC_{OB}}{dt} \tag{20.2}$$

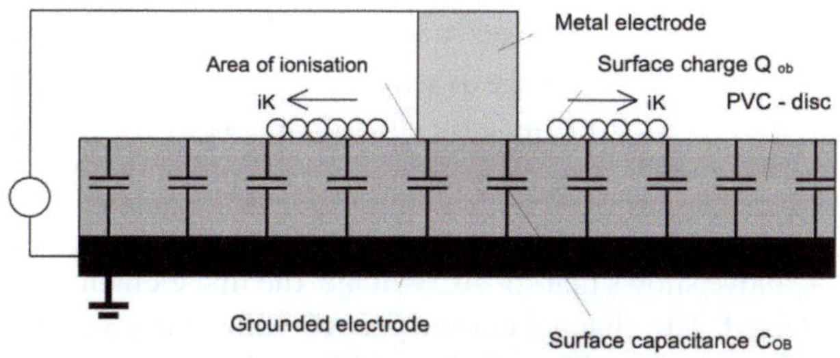

Principle arrangement according to Toepler to explain the physical processes of a surface discharge.

Typical surface discharges and surface flashover on a PVC disc at applied voltage of 30 kV rms. This recording was made with AC voltage. With a surge voltage, a single streamer would occur.

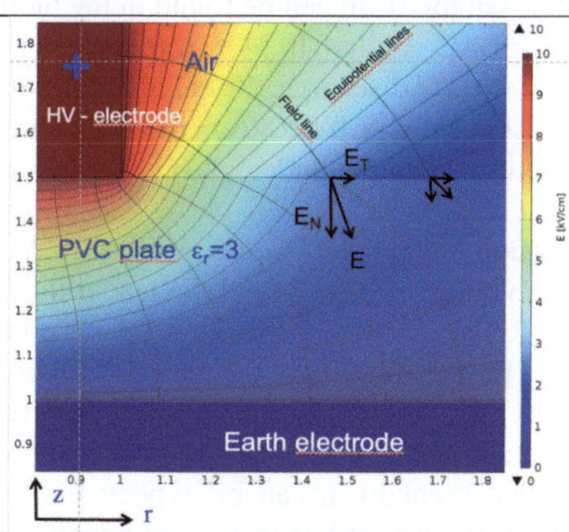

Associated field plot of the electric field strength with equipotential surfaces and lines in an r-z-coordinate system.
EN Normal component of the electric field strength on the surface.
ET Tangential component of the electric field strength on the surface E

Fig. 20.3 Example for an explanation of the surface discharge

ε_0	Permittivität of vacuum
ε_r	Permittivität of XLPE
A	Charged Arera
Q_{OB}	Charge on surface
C_{OB}	Capacity of the charged surface
U	Applied voltage

The equation above shows that for AC voltage, the first element on the right-hand side can be neglected. The channel current for AC voltage is given by the change in surface capacitance over time. This is to be understood in such a way that the charge-occupied surface changes in time and therefore the surface capacitance, which is so determined by definition, changes. With AC voltage, the channel current is therefore dependent on the specific surface capacitance. For impulse voltages, the temporal change of the applied voltage is decisive for the level of the channel current and the following then applies

$$i_K = \frac{\varepsilon_0 \varepsilon_r}{s} \frac{dU}{dt} \tag{20.3}$$

Above a critical channel current strength, thermo-ionisation sets in and the surface sparks are produced, which are noticeable by an intensive glow and strong noise development. The discharge is described by a falling current–voltage characteristic of the arc and the strength of the section drops to specific values of a few kV/cm.

In the literature, information on gliding discharges with AC voltage is documented in [27, 36, 37]. For the formation of surface discharges with impulse voltage, only data in the range of a few cm to 10 cm can be found in the literature [40, 41], which confirms the basic laws for AC voltage. In [41], reference is made to the extremely short flashover time for a surface discharge. No reliable formulaic data can be found in the literature for the flashover range from 80 cm to a few metres. Surface discharges on the surface of an insulating material under a layer of water have been investigated in [42].

As a general orientation, the relationships applicable to AC voltage can be used.

The corona inception voltage stress U_K and the leader inception voltage U_G can be calculated according to [27] for dry insulating layers as a function of the applied stress, the thickness s and the permittivity of the material. Figures 20.4 and 20.5 show the corresponding graphs for materials with a relative permittivity of 2–5.

The corona inception voltage is not equal to the breakdown voltage. However, experience has shown that breakdown can be expected with a small increase in voltage. The specific dielectric strength is then 1–2 kV/cm.

$$U_K = 11,452 * (s/\varepsilon)^{-0,45} \tag{20.4}$$

$$U_G = 1,91 * 10^{-4} * (c')^{-0,44} \tag{20.5}$$

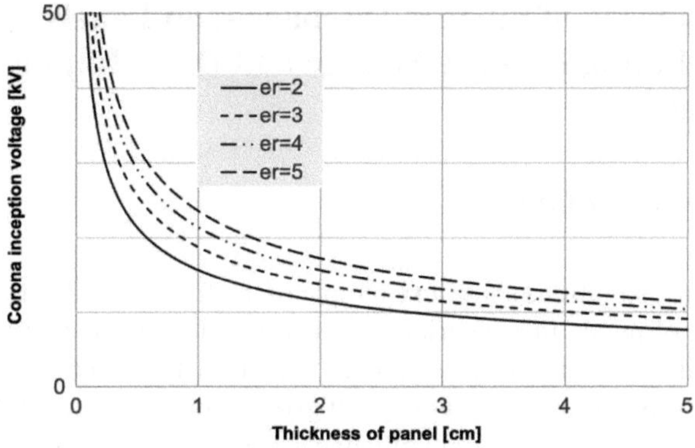

Fig. 20.4 Corona inception voltage of a surface discharge of insulating panels, calculated according to data in [24], parameter er is the permittivity

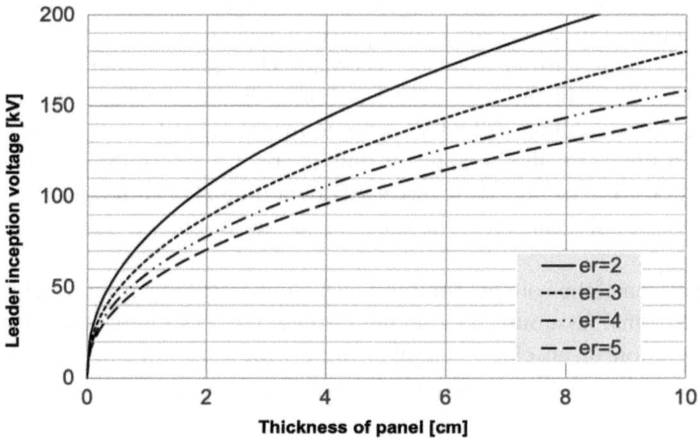

Fig. 20.5 Leader inception voltage of a surface discharge of insulating panels calculated according to data in [24], parameter er is the permittivity

$c' = 8{,}85*10^{-12}$ [F]*ε_r/(100 cm *s [cm])) Eq. 20.6.
ε_r: Relative Permittivity, for asphalt 2,4…3,3
s: Thickness of insulation.

20.3 Comparison of Surface Discharges on Panels at Lightning Impulse Voltage and Delta-Impulse Voltage

20.3.1 Tests on Panels

The following investigations show an example of the formation of surface discharges on a plate made of insulating material. The investigations were carried out on a clay tile and a granite plate in the arrangement shown in Fig. 20.6. This arrangement differs from the typical surface arrangement given in the literature in that the back of the plate is not completely covered with an earthed plate. Nevertheless, pronounced surface discharges can be seen here.

First, the lightning impulse voltage 1.2/50 according to Fig. 20.6, which is known from test standards, was applied. To compare this with the induced voltage occurring at the down-conductor during a lightning strike with a negative follow-up flash, this was simulated by a so-called delta pulse, which is shown in Fig. 20.7. In both cases, the discharge current was also recorded. The results of the investigation are shown in Fig. 20.9. In the case of the lightning impulse voltage, for example, a smaller breakdown voltage (72.8 kV) is found for the clay tile than for the pulse load (120 kV). In Fig. 20.8, the breakdown time is about 5 μs. In the case of the pulse voltage and clay tile in Fig. 20.9, the breakdown time is only about 0.3 μs (start of the current rise until flashover). The speed of surface discharges is reported in Heinz et al. [33].

A comparison of the voltage–time surfaces shows a striking difference between the two voltage forms. Assuming a radial field strength of 25 kV/cm, which is necessary for the formation of an electron avalanche, a necessary voltage of 15 kV results from the numerical field calculation of the arrangement in Fig. 20.10. Calculating

Fig. 20.6 Laboratory set-up for the investigation of surface discharges on a clay tile. The electrode resting on the tile is painted black to avoid reflections during photography. The back of the clay tile is not metallised, only the counter-electrode is on the surface

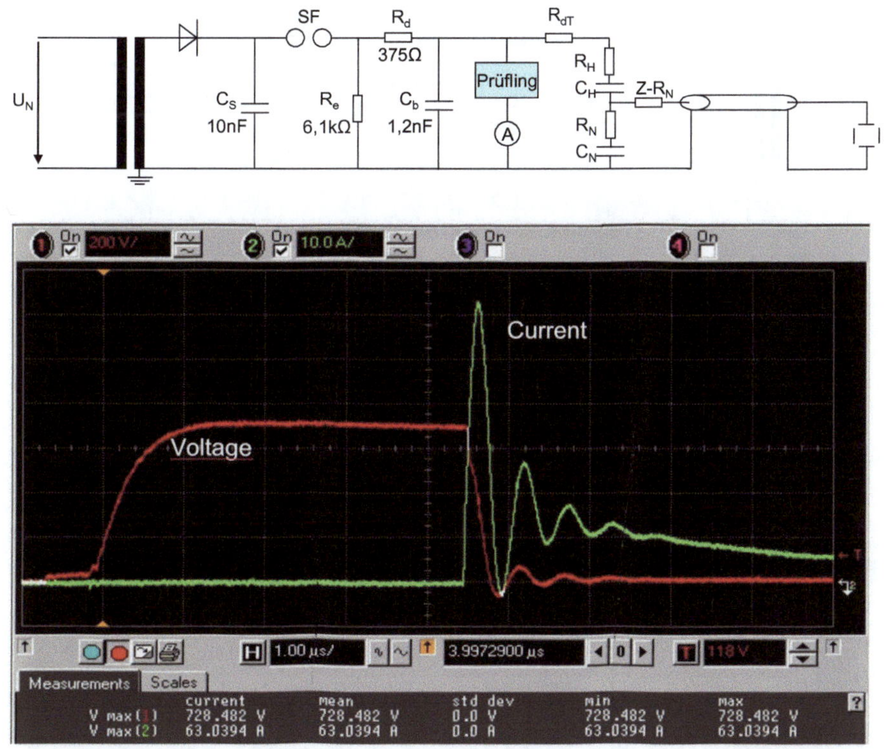

Fig. 20.7 Circuit configuration and oscillogram at lightning impulse voltage 1.2/50. The voltage scale is to be multiplied by 100

the voltage–time area above 15 kV up to the flashover of the arrangement results in the following values:

1.2/50, Fig. 20.7: 200 kVμs = 0.2 Vs

Pulse, Fig. 20.8: 24 kVμs = 0.024 Vs.

An area-time law cannot be derived from these results. One possible explanation for the striking difference is the formation of gliding sparks, which are more intense with pulse stress due to the larger dU/dt. The glide sparks also glow more intensively with pulse stress, see Fig. 20.9. On the other hand, the arrangement according to Fig. 20.6 is characterised by the fact that the back of the tile is not metallised. However, this experimental finding must be taken into account when developing a test method for isolated conductors. However, the formation of surface discharges on a coaxial system must be calculated.

As a conclusion, it can be stated that surface discharges on an insulating plate leads to flashover within a few 100 ns when a voltage is applied that corresponds to the form of an induced voltage on a down-conductor.

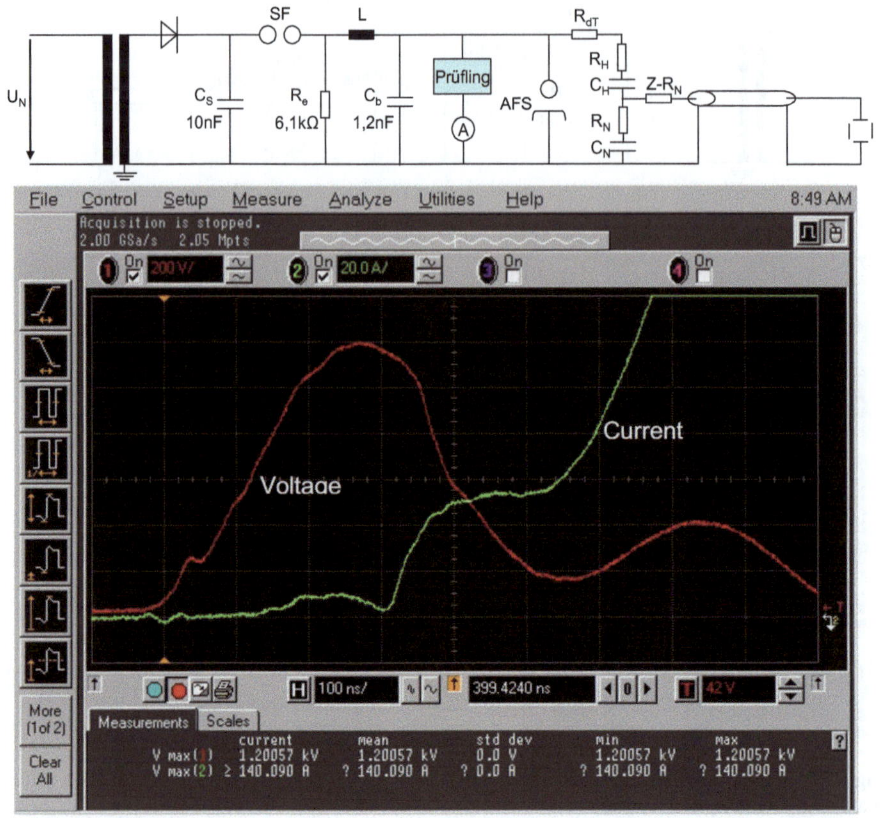

Fig. 20.8 Circuit configuration and oscillogram at Delta-impulse "0,25/100". The voltage scale is to be multiplied by 100. *Note* delta-impulse means: Voltage shape is the derivative di/dt of a current having a shape 0,25/100 according to IEC 62305-1

20.3.2 *Field Calculation for the Test Arrangement Used for Tests*

For the clay tile shown in Fig. 20.6, a numerical field calculation with the model arrangement according to Fig. 20.10 is carried out in the following. The material data are shown in Table 20.1. The field plot is shown in Fig. 20.11 and enlarged in Fig. 20.12. As expected, the high field strength at the edge of the metal electrode can be seen. The field strength values were evaluated on the surface of the insulating material, see Fig. 20.13.

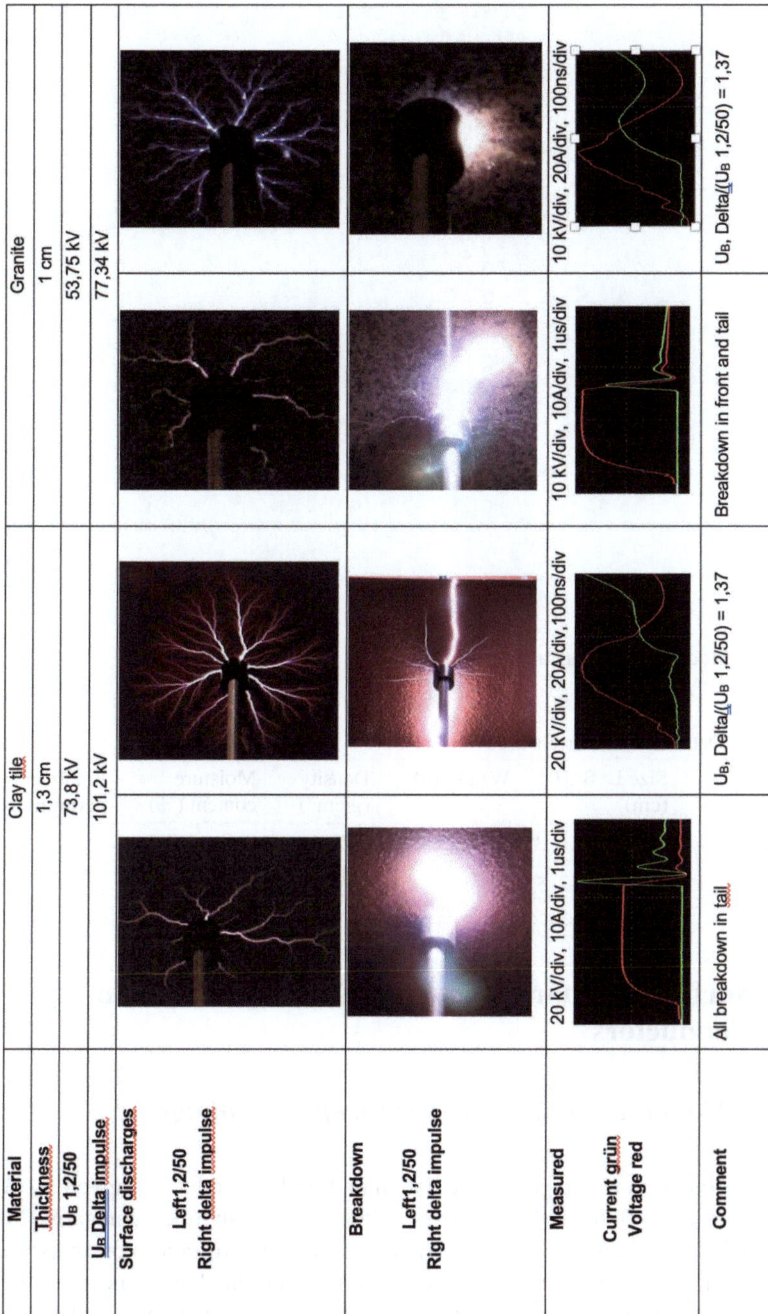

Fig. 20.9 Comparison of results. *Source* Diplomarbeiten Herr Balkenohl und Herr Koppe, Fachhochschule Südwestfalen 2008

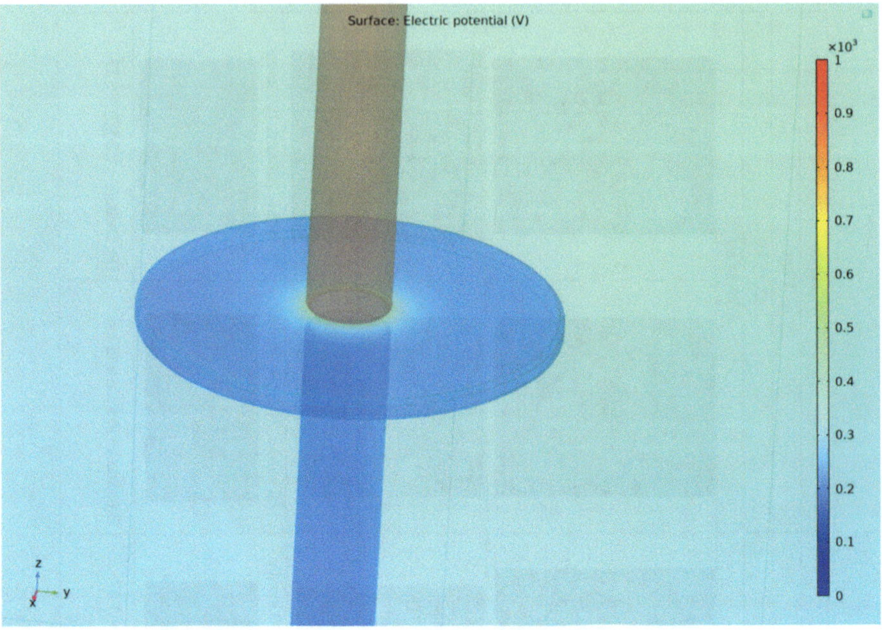

Fig. 20.10 Model for field calculation

Table 20.1 Data of the materials used

Device under test	Size L*B*H (cm)	Weight (g)	Density (g/cm^3)	Moisture content (%)	Relative permittivity ε_r
Clay tile	24 × 24 × 1.3	1713.35	2.29	0	7.25
Granite	30.5 × 30.5 × 1	2576	2.77	0.5	7.16

20.4 Surface Discharges on Coaxial Insulating Down Conductors

20.4.1 Estimation of the Leader Inception Voltage

The leader inception voltage can be calculated with the relations given by Böhme [24]. According to Böhme, the flashover voltage of smooth tubes at negative impulse voltage 1.2/50 is about 35% greater than at alternating voltage. For an estimation, the flashover voltage can be set approximately equal to the leader inception voltage, whereby a 1.35-fold value is then to be applied for negative surge voltage 1.2/50. This relationship is shown in Fig. 20.14. This relationship is shown in Fig. 20.14, where a factor of 1.35 was introduced in Eq. 19.2 and thus Eq. 20.1 applies. The

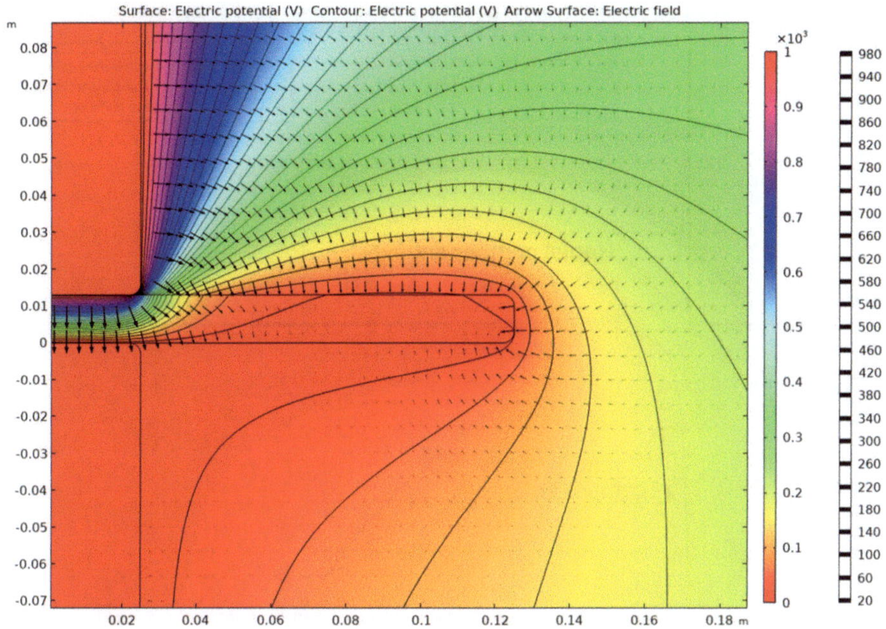

Fig. 20.11 Field plot with equipotential lines and areas as well as vectors of the electric field strength in logarithmic representation for an applied voltage of 1000 V

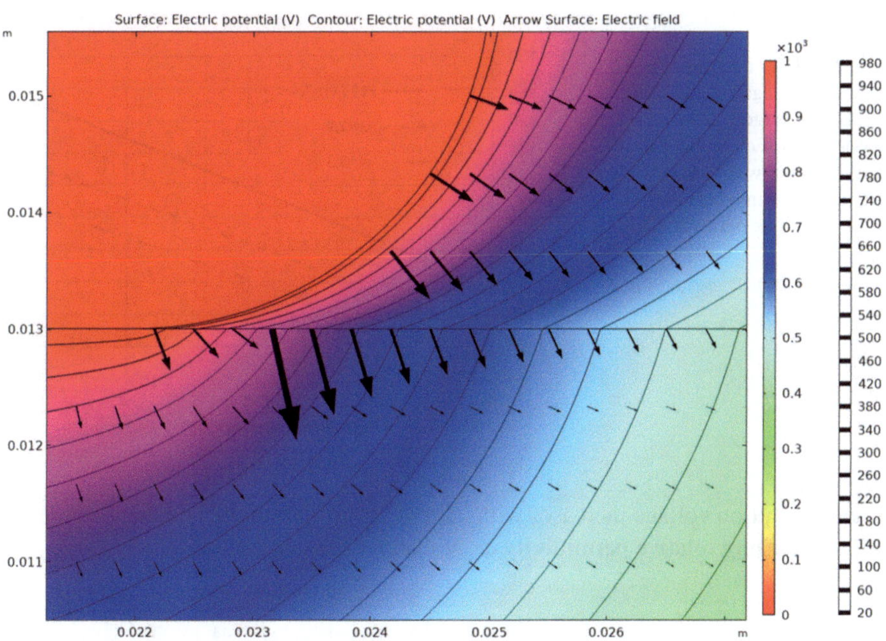

Fig. 20.12 Higher resolution of Fig. 20.11

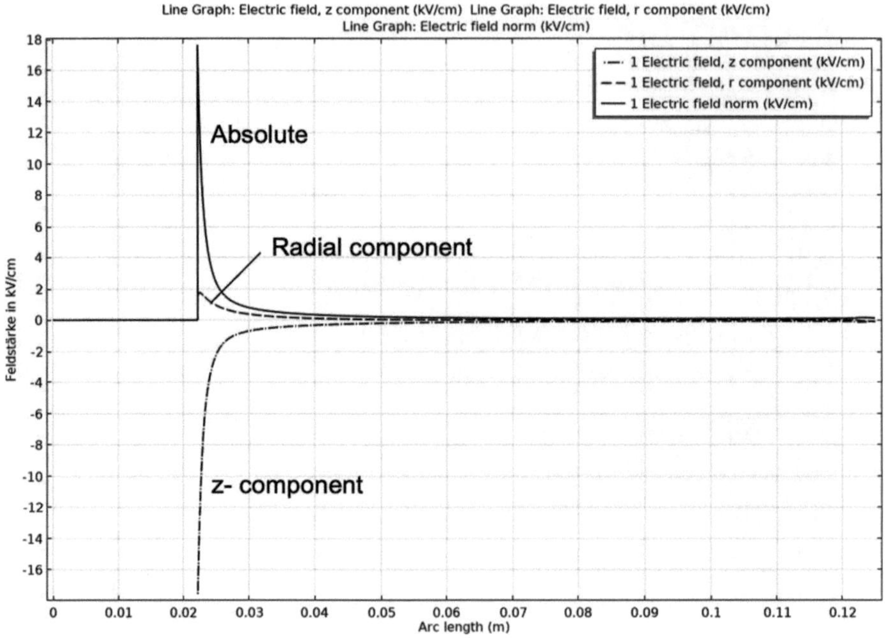

Fig. 20.13 Components of the electric field strength on the surface of the clay tile. Illustration for an applied voltage of 1000 V

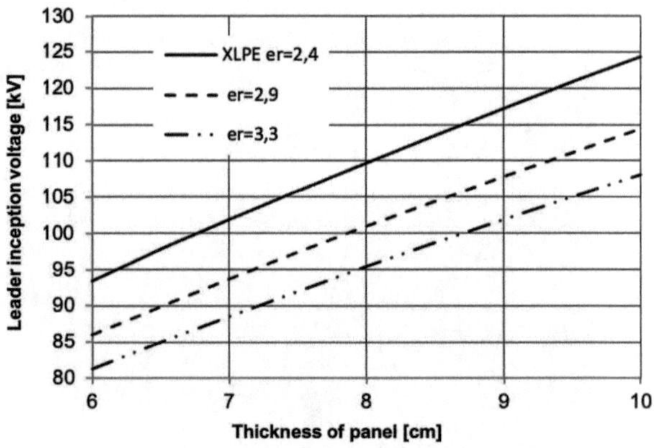

Fig. 20.14 Leader inception voltage for negative lightning impulse voltage 1.2/50 according to Eq. 20.6 for a coaxial cable with 8 mm diameter and insulation of thickness s. er: permittivity

leader inception voltage increases with increasing insulation thickness and decreases with increasing relative permittivity er. No values are known for delta pulses.

$$U_{G,1,2/50} = 1,35 * 1,91 * 10^{-4} * (c')^{-0,44} \qquad (20.6)$$

20.4.2 Laboratory Test on a Conductor with XLPE Insulation

The leader inception voltage has been investigated on an insulated conductor in the laboratory. A commercially available RG 218/U coaxial cable was used for this purpose, with the shield of this coaxial cable removed. A sample was arranged as a ring according to Fig. 20.15 and the test voltage was connected to the connected ends. A current measurement was also used to check the flashover. A high-resolution camera was used to register the discharge development during a surge voltage stress. The voltage was generated with a Marx generator with short-circuited internal damping resistors and an external coil. The shape of the oscillating surge voltage is shown in Fig. 20.17. This voltage shape still differs in the pulse width by a factor of 20 compared to the pulse width of a calculated voltage pulse according to Fig. 17.14. The expected gliding spark insertion voltage is 95 kV according to Fig. 20.14.

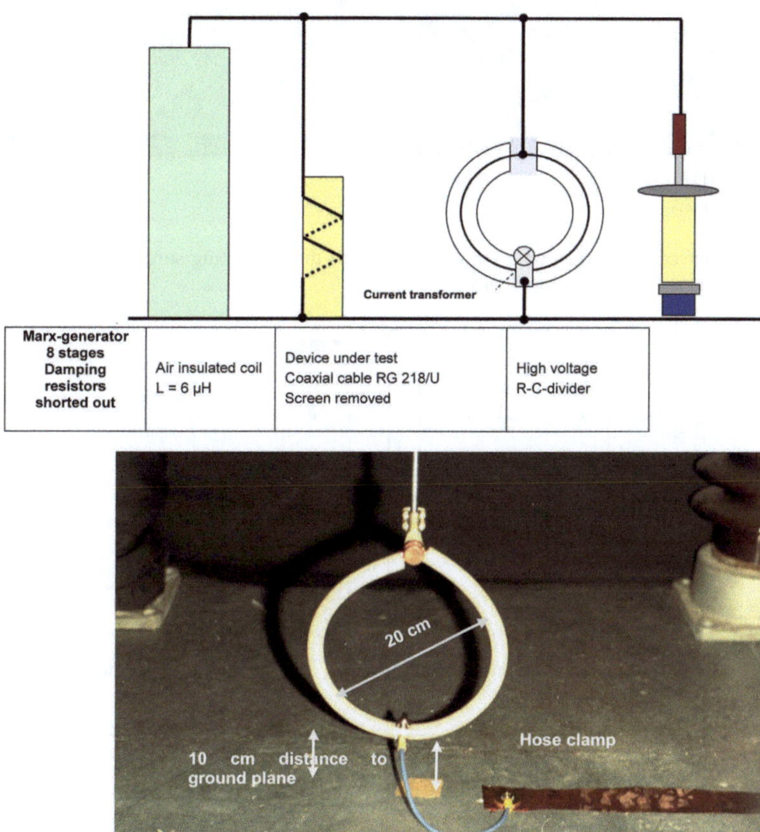

Fig. 20.15 Test setup for the surface discharge test on a coaxial cable RG 218/U (the metallic shield has been removed). Conductor diameter: 4.95 mm. Outer diameter of the insulation: 17.3 mm, wall thickness 6.2 mm

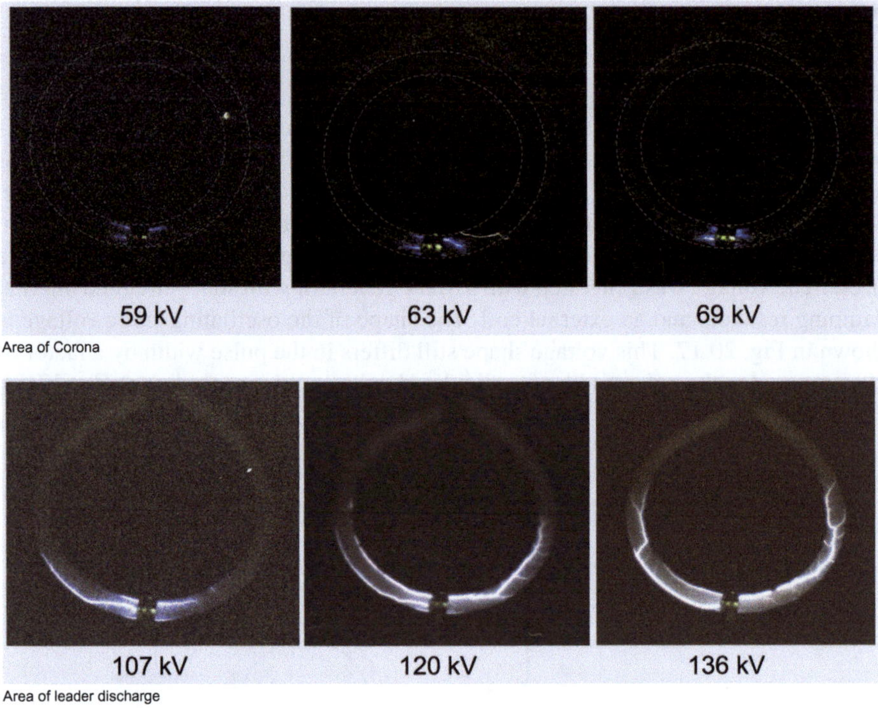

Fig. 20.16 Surface discharges with increasing level of the oscillating surge voltage on a coaxial cable RG 218/U according to Fig. 20.15

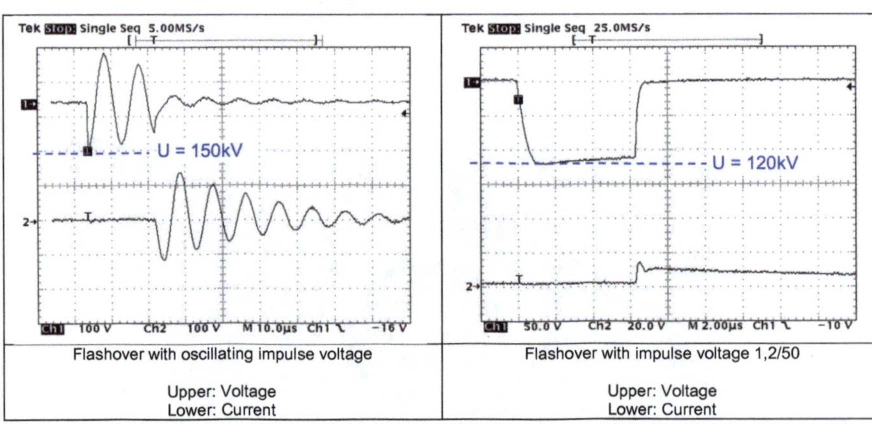

Fig. 20.17 Measured voltage and current during tests

The results in Fig. 20.16 show a surface corona already from a surge voltage level of 59 kV. Surface sparks were observed from a height of 107 kV. At a voltage level of 136 kV, the arrangement began to flash over. However, this voltage level is within the range of the voltages given in Table 13.2. This shows that an insulated arrester used in a down-conductor must be tested in a type test to ensure that it is free of surface discharges. In this case, it will only be possible with a great deal of effort to generate such a rapidly rising and falling surge voltage according to Fig. 19.4. Further research is therefore necessary (Fig. 20.17).

Chapter 21
Annex A: A Contribution to the Limitation of Step Voltages

21.1 State of Art

The installation of a lightning protection system according to the state of the art [1] includes compliance with the requirements specified in Sect. 21.8 [1]. This specifies the protective measures for maintaining the step and touch voltage:

- *Under certain conditions, the proximity to the down conductors can be life-threatening even if the LPS was designed and installed in accordance with the above requirements.*
- *The danger is reduced to an acceptable level if one of the following conditions is fulfilled:*

  (A) *Under normal operating conditions there are no persons within a radius of 3 m from the discharges*
  (B) *A system of at least ten down conductors corresponding to 5.3.5 is available. N5)*
  (C) The contact resistance of the superficial soil layer is not less than 100 kOhm

- Note A layer of insulating material, e.g. asphalt 5 cm thick (or a layer of gravel 15 cm thick), generally reduces the risk to an acceptable level.
- If none of these conditions is met, the following protective measures must be taken to avoid injury to persons due to step voltage.
  - Potential equalisation by means of a meshed earthing system
  - Barriers and/or warnings to reduce the probability of entering the hazardous area within 3 m of the discharge

This paper is the translation of "Ein Beitrag zur Begrenzung der Schrittspannung durch isolierende Schichten "VDE/ABB-Blitzschutztagung, Neu Ulm, 2015.

J. Meppelink, *The Lightning Rod as a Danger*,
https://doi.org/10.1007/978-3-031-30434-7_21

The state of the art has changed since the publication of [1]. In addition to the above protective measures, further questions arise in practice:

- What step voltage occurs in an asphalt layer covered with rain-water?
- Do sliding discharges occur?
- Does soil dehydration occur under an asphalt layer?
- How does the ionisation in the soil affect the step voltage?
- How does a soaked gravel layer behave?
- How does the specific resistance of a gravel layer change due to environmental influences?
- What ageing is to be expected with asphalt?
- What effects occur at the edges of an asphalt layer?

The paper provides answers to these questions using the example of a shelter typical for mountain refuges in the event of a lightning strike of 100 kA of form 10/350. This covers the worst case [2, 3]. The calculation of other surge current forms is therefore not necessary.

21.2 Insulation Material

Asphalt [4] is a naturally occurring or technically produced mixture of bitumen and coarse and fine aggregates and, where applicable, additives. Mastic asphalt is waterproof and practically impermeable to gas and water vapor. Asphalt is stable over many years. Therefore, the soil beneath a larger asphalt layer will dry out to a certain extent. Asphalt has a dielectric strength of 10…60 kV/mm and a permittivity of 2.4…3.3 [5]. The specific resistance of asphalt and gravel is described in detail in [6]. Crushed rock is shown in [7] with no electrical properties given. In view of the wide range of different materials, it is recommended to measure the specific resistance of the intended material [8]. According to [6], it must be taken into account that, over time, the spaces in between are filled by loose soil underneath the gravel layer and dust is deposited from the air. These effects reduce the specific resistance (Table 21.1).

21.3 Calculation with XGS-Lab (Grounding System Analysis)

As described in detail in [2, 3], the worst-case scenario for the step voltage is the pulse 10/350, which is simulated with a sinusoidal current of 25 kHz in the frequency range. This approach is justified for physical considerations in linear systems with purely ohmic coupling, which are connected to the peak value of the surge current,

Table 21.1 Specific resistance of materials for insulating layers [6]

Material	Specific resistance (Ohmm)	
	Dry	Wet
Asphalt	$2*10^6 \ldots 30*10^6$	$10.000\ldots 0.2*10^6$[1]
Beton Concrete	$10^6 \ldots 10^9$	21..... 200
Crushed granite with fine	$140*10^6$	1300 With groundwater 45
0.04 m Crushed granite with fine	4000	1200 With rainwater 100
0.025 m bis 0.1 m Washed granite Similar to **Gravel**	$1{,}5*10^6$ ...... $3*10^6$	5000 With rainwater 100

and was verified by comparative calculations with lattice circuit models and with time-domain calculations with XGS-LAB.

The propagation speed of a transverse electromagnetic wave (TEM wave) depends on the specific earth resistance and the frequency of the excitation current. For a quasi-stationary calculation with a fixed frequency of 25 kHz, the dimension of the grounding system is limited to 10% of the wavelength. Figure 21.1 shows the dependence of the /10 wavelength on the specific earth resistance. Thus, the permissible value for a calculation with a specific earth resistance of 100 Ohmm is about 30 m and is therefore completely sufficient for the calculation presented here on an earthing system with dimensions of 5 m × 5 m. However, if the dimensions of the earthing system are larger than the corresponding /10 wavelength, unsteady conditions exist, whereby travelling waves occur in the earthing system. In these cases, a calculation with XGS-LAB in the time domain is recommended. The standardized pulse forms [9] can be used.

The program takes the magnetic coupling of conductor loops into account and allows the use of conductors above the earth's surface and insulating layers on the earth's surface. Thus, entire lightning protection systems can be modelled. The earth can be modelled by multilayers and multiarea. The specific earth resistance and permittivity are modelled depending on the frequency. The specific earth resistance is usually not precisely known and depends on the temperature, soil moisture and salinity of the soil [6]. For the permissible step and touch voltage, permissible electric and magnetic fields, limit values can be entered, the exceeding of which is then visualized in the solution. Electric and magnetic fields can be calculated as absolute values and as components, but no vectors. XGS-LAB allows a comfortable representation of the no-load step voltage and the loaded step voltage for a body resistance of 1 kOhm. The permissible step voltage is determined according to the specifications of [10] to 25 kV 10/350 corresponding to a specific energy of $W/R = 0.156\ A^2s$ and is

[1] It can be assumed that the conductivity is reduced by the water film on the surface.

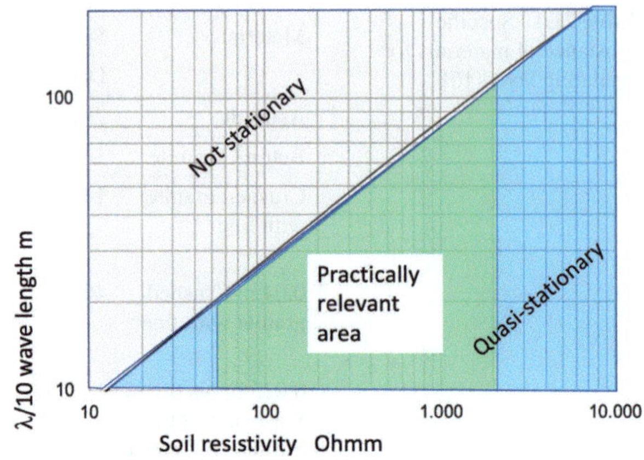

Fig. 21.1 l/10 Wavelength as a function of the soil resistivity for a frequency of 25 kHz corresponding to a pulse current 10/350

also intended for the new version of IEC 62305-3. A value of 2 kV was proposed for the touch voltage [11]. XGS-LAB enables a comfortable presentation of the results.

21.4 Calculation with Comsol Multiphysics

Comsol Multiphysics is a program package for the general solution of partial differential equations with finite elements. The program is particularly suitable for the visualization of electromagnetic fields with vectors. Grounding systems can be calculated whereas the step voltage under load conditions of 1 kOhm[2] cannot be displayed without great effort. In this paper the program is used as a supplement to XGS-LAB for the following questions:

- Testing of sliding discharges on insulating surfaces
- Step voltage at edges of limited asphalt layers.

21.5 Calculation Including Soil Ionisation

Ionisation of the soil at electrodes is not considered in XGS-LAB. Therefore, a possibility to insert ionisation was looked for.

According to [12, 13], the surge earthing resistance R_{ST} for earthing systems with dimensions of 3–30 m and surge currents with end times >0.2 μs can be calculated with Eq. 21.1 can be displayed. For surge currents 10/350 the surge factor is: a = 1; [2, 3]. This results in a reduction factor K for the consideration of the ionisation in the soil according to Eq. 21.3, which is shown in Fig. 21.2 for a surge current 10/350 with 100 kA peak value. The peak value of the step voltage is proportional to

[2] The resistance of a human body is considered as 1 kOhm.

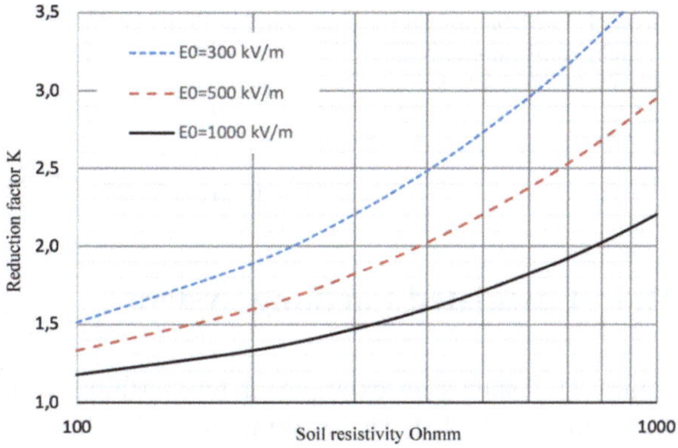

Fig. 21.2 Reduction factor K according to Eq. 21.3 für I = 100 kA 10/350

the surge earthing resistance R_{ST}. For a calculation in XGS-LAB with ionisation the permissible step and touch voltage is multiplied by the reduction factor K according to Tables 21.2 and 21.3. The calculation then shows the limits at which the values of 25 or 2 kV are reached by the effect of ionisation.

$$R_{ST} = R_A * A_i \tag{21.1}$$

$$A_i = \frac{1}{\sqrt{1 + \frac{I}{I_G}}} + \alpha - 1; I_G = \frac{E_G * \rho_E}{2\pi * R_A^2} \tag{21.2}$$

$$K = \frac{1}{A_i} = \sqrt{1 + \frac{I}{I_G}} \tag{21.3}$$

Ai: Coefficient for ionisation and travelling waves.
I: Surge current in kA (in the range 0–100 kA).
I_G*:* Limit current according to.
ρ_E*:* Specific earth resistance in Ohmm.
(In the range 10–1000 Ohmm).
E_C*:* Limit field strength (300–1000 kV/m).
R_A*:* Stationary propagation resistance = *7.85 Ω*.
α: Surge factor, which takes into account travelling waves in the earth.
K: Reduction factor for ionisation.

Table 21.2 Consideration of ionisation. Indices mean: (P: Prospective; S: Step; T: Touch)

Ionisation of soil	Permissible step voltage	Permissible touch voltage
No ionisation	U_{SP} = 25 kV	U_{TP} = 2 kV
With ionisation	U_{SP}*K: 25 kV*K	U_{TP}*K = 2 kV*K

Table 21.3 Reduction factor K to take ionisation into account. I = 100 kA; R_A = 7,85 Ω

	Limit field strength E_G for Ionisation		
Soil resistivity (kV/m)	300	500	1.000
r_E = 100 Ohmm	1.513	1.332	1.178
r_E = 1000 Ohmm	3.728	2.956	2.207

21.6 Calculated Earthing System

Figure 21.3 shows an example of the earthing system of a shelter calculated in accordance with [15]. The earthing grid with the dimensions 5 × 5 × 0.25 m with stainless steel rods of 5 mm diameter is arranged at a depth of 25 cm. The interception system with the two down conductors is modelled from 8 mm stainless steel rods. Insulating layers are indicated in the further studies.

Measurements of the specific earth resistance have shown fluctuations of up to 6 times the measured maximum value [14]. Before planning an earth-termination system, the earth resistivity should be determined by measurement. Furthermore, seasonal fluctuations, especially low temperatures, should be taken into account.

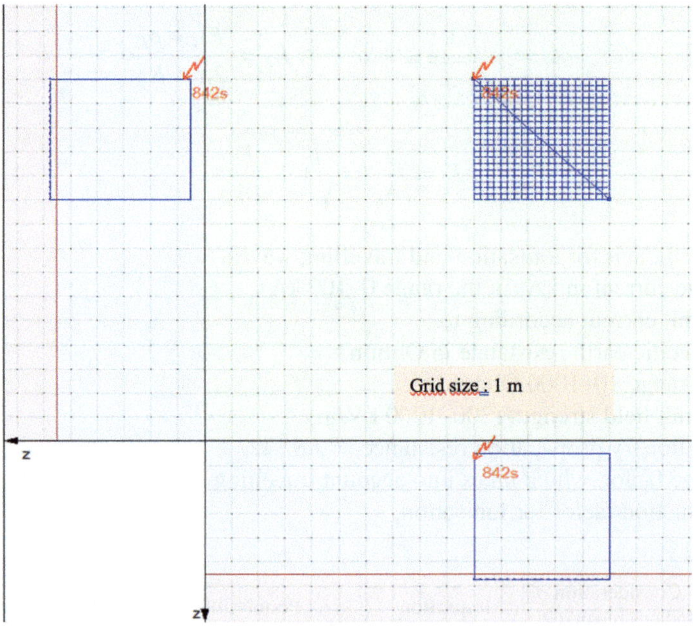

Fig. 21.3 Calculated exemplary earthing system in isometric view (detail)

All calculations were carried out with the equivalent frequency of 25 kHz for the pulse shape 10/350 with a value of 100 kA according to protection class III/IV. The following variables were considered:

- Specific earth resistance
- Specific resistance of insulating layers
- Limit field strength of the soil (ionisation).

21.7 Convention for the Results of the Calculation

	Result of the calculation for the limit values. Permissible touch voltage: 2 kV 10/350; Permissible step voltage: 25 kV 10/350
Green	Step- and touch voltage are not dangerous
Yellow	Touch voltage is potentially high1) Step voltage is not dangerous
REd	Step- and touch voltage are not dangerous 1)
	1) People whose central axis is in the red or yellow area are at danger.

21.8 Step Voltage Without Insulating Layer

The calculation in Fig. 21.4 shows that even assuming good soil with $r_E = 100$ Ohmm at the edges and outside of the grid there is danger, even taking into account the ionisation in the soil.

The following measures can be taken to avoid the danger of excessive step voltage at the edges:

- Potential control by additional earth conductors
- Dry 5 cm high asphalt layer, which covers at least the red area for $\rho_E = 1{,}000$ Ohmm
- Dry gravel layer 15 cm high.

An insulating layer (asphalt with the lowest value according to Table 21.1) is calculated in XGS-LAB as a layer covering the entire ground. Figure 21.5 shows the effect of the asphalt or gravel layer and the influence of the specific earth resistance. Asphalt and gravel reduce the step voltage.

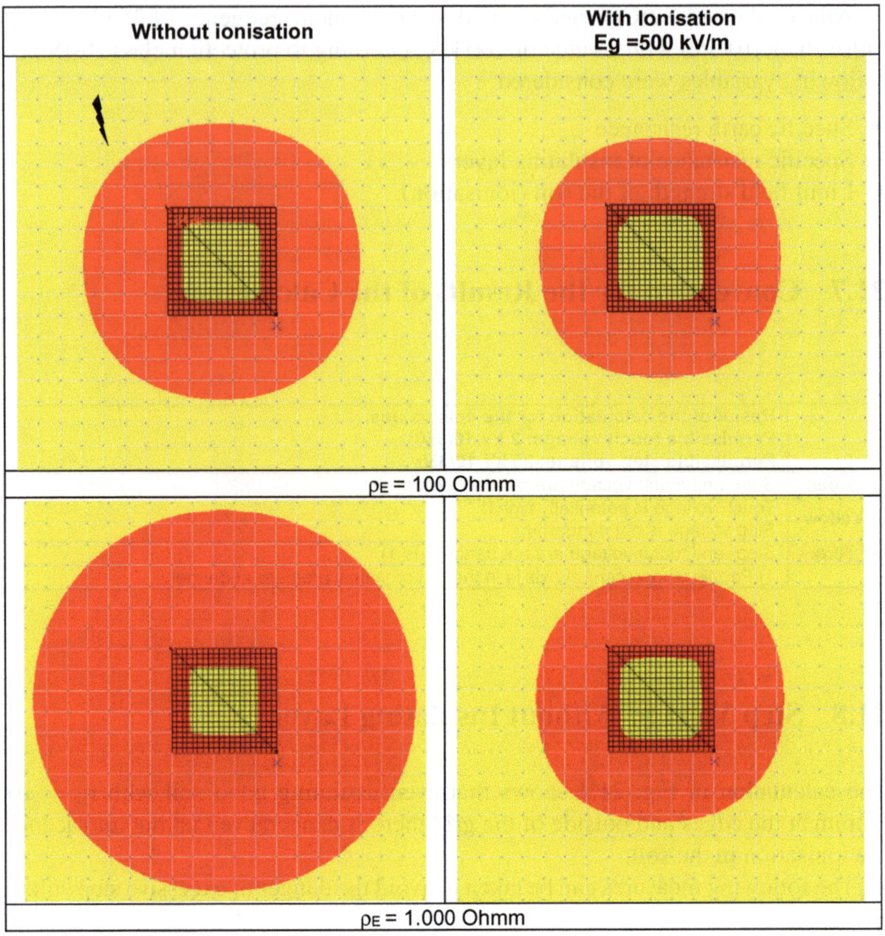

Fig. 21.4 Step and touch voltage of the earthing system. Grid size: 1 m

21.9 Step Voltage with a Dry, Insulating Layer of 5 cm Asphalt or 15 cm Gravel

An insulating layer (asphalt with the lowest value according to Table 21.1) is calculated in XGS-LAB as a layer covering the entire ground. Figure 21.5 shows the effect of the asphalt or gravel layer and the influence of the soil resistivity. Asphalt and gravel reduce the step voltage.

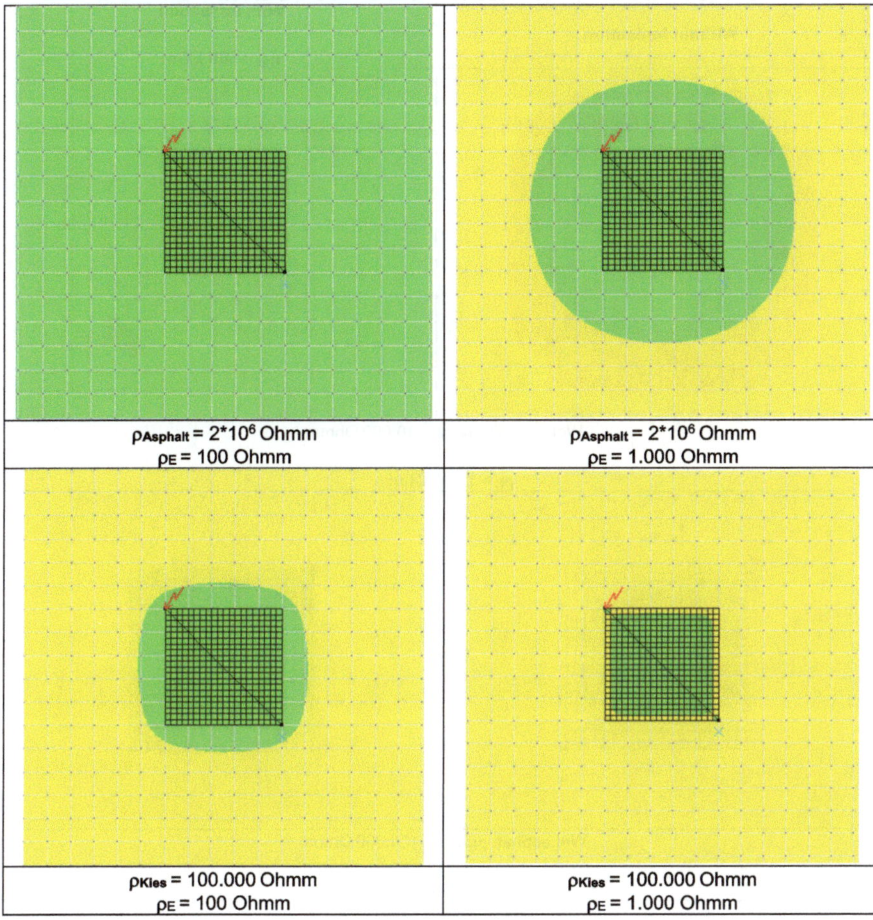

Fig. 21.5 Step and touch voltage for dry iso-liquid layer of 5 cm asphalt or 15 cm gravel

21.9.1 Step Voltage with a Wet Layer of Asphalt

With the values for wet asphalt according to Table 21.1 and the assumption that the soil beneath a larger asphalt layer has dried out, all following calculations are carried out as worst case calculations for a specific earth resistance of $\rho_E = 1{,}000$ Ohmm. Wet asphalt in this case means asphalt with a specific volume resistance, i.e. asphalt without a surface water layer. Figure 21.6 shows that asphalt with $\rho_{asphalt} = 10.000$ Ohmm is not sufficient. Assuming ionisation in the ground, even a layer of asphalt $= 10{,}000$ Ohmm is sufficient, except for the edges at the leads. With asphalt with $\rho_{asphalt} = 100.000$ Ohmm there is no danger above the grid.

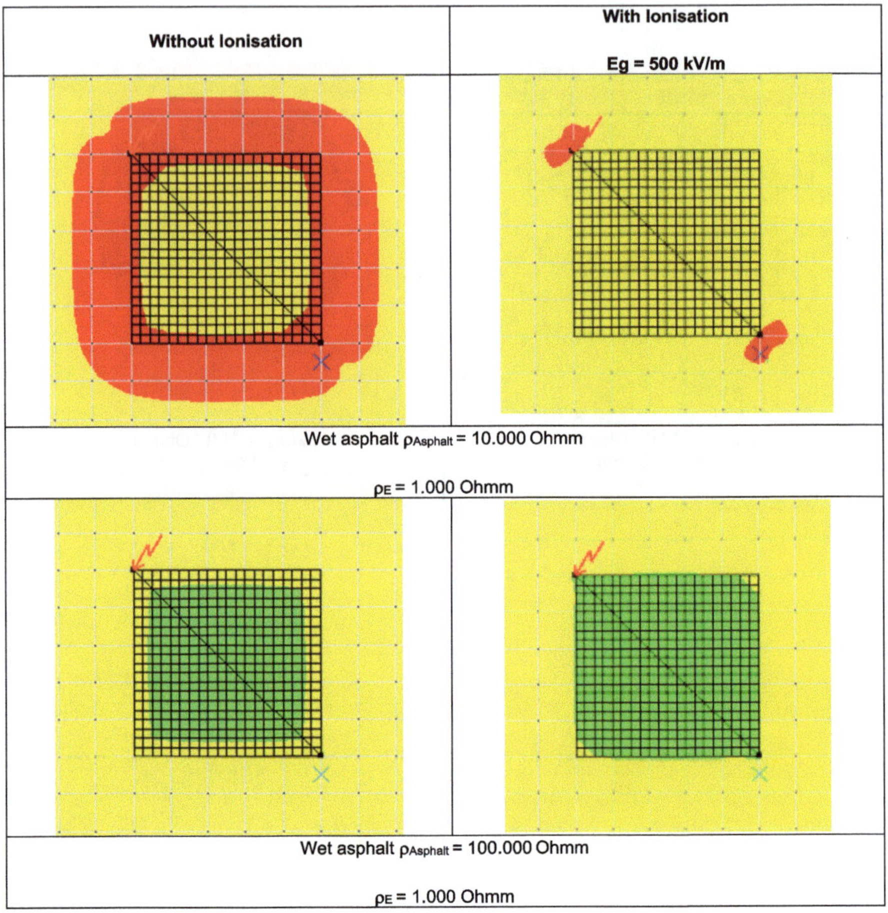

Fig. 21.6 Step and touch voltage for an insulating layer of 5 cm wet asphalt

21.9.2 Step Voltage with a Wet Layer of Gravel

A wet gravel layer according to Table 21.1 and Fig. 21.7 is uncritical for the step voltage up to $\rho_E = 300$ Ohmm. The touch voltage is potentially dangerous.

21.9.3 Step Voltage for a 10 cm High Rain Water Layer on an Insulating Layer of 5 cm Asphalt

Although wet asphalt has already been investigated in the above section, a 10 cm high layer of rainwater with $\rho_W = 45$ Ohmm on a dry, 5 cm high asphalt layer with a

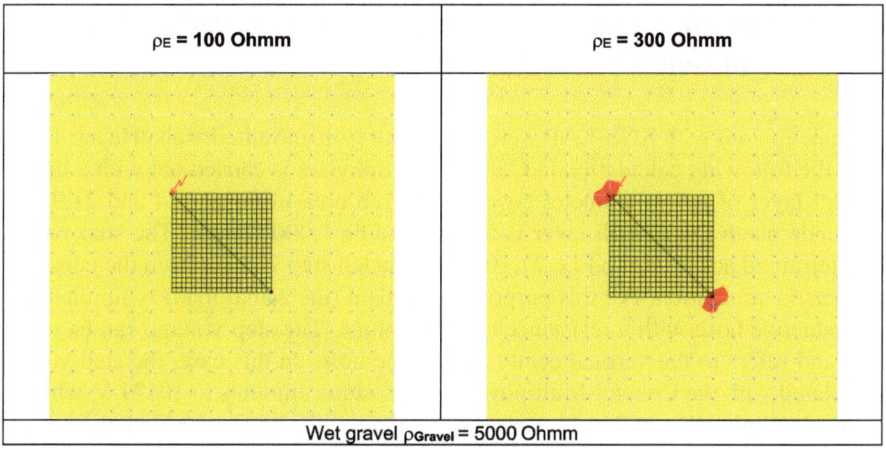

Fig. 21.7 Step and touch voltage in a wet gravel layer 15 cm high

resistivity of $\rho_{\text{asphalt}} = 2*10^6$ Ohmm over a dry layer of earth with $\rho_E = 1{,}000$ Ohmm is calculated below. The step and touch voltages are calculated with the XGS-LAB program on the water surface and are too high according to Fig. 21.8. Heavy rain is therefore a hazard despite the grid and highly insulating asphalt. According to Sect. 6.8 this also applies to a 2 cm high water surface.

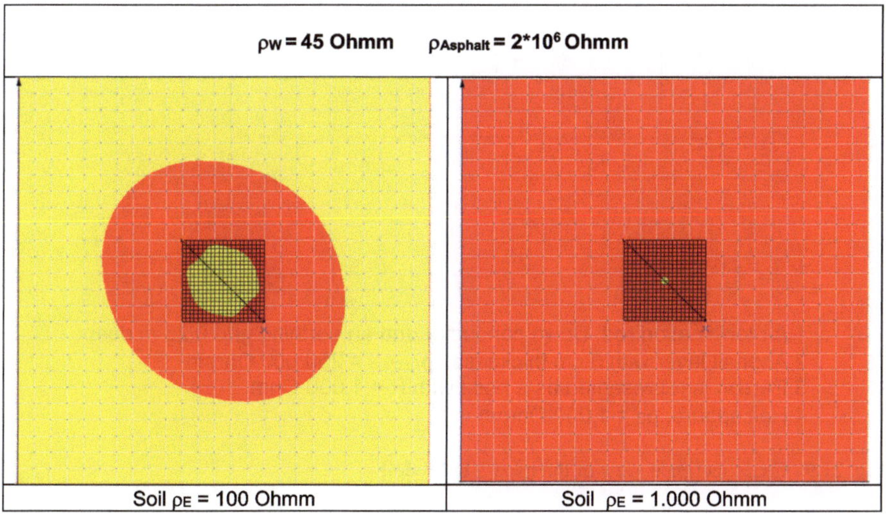

Fig. 21.8 Step and touch voltage with a water layer of 10 cm height on asphalt of 5 cm

21.9.4 Calculation of the Edge Effect on a Dry Asphalt Layer with Comsol-Multiphysics

The calculations with XGS-LAB were carried out with unlimited asphalt layer. Therefore, the following calculation in Comsol-Multiphysics is carried out with a limited asphalt layer of 3.5 m diameter covering the red area in Figs. 21.4 and 21.9. The soil underneath the asphalt layer is assumed to be 1,000 Ohmm. The step voltage is determined according to Fig. 21.10 as the loaded step voltage from the current in the person simulation. For this purpose the person (the actual load) is simulated by a conductive hose with a resistance of 1,000 ohms. The step voltage can be evaluated and refers to the vertical centre axis of the hose. In this case, the step voltage calculated with the Comsol-Multiphysics programme amounts to 10121 V, which is below the limit value of 25 kV. A person standing with one foot on asphalt and with the other foot in the red zone is safe.

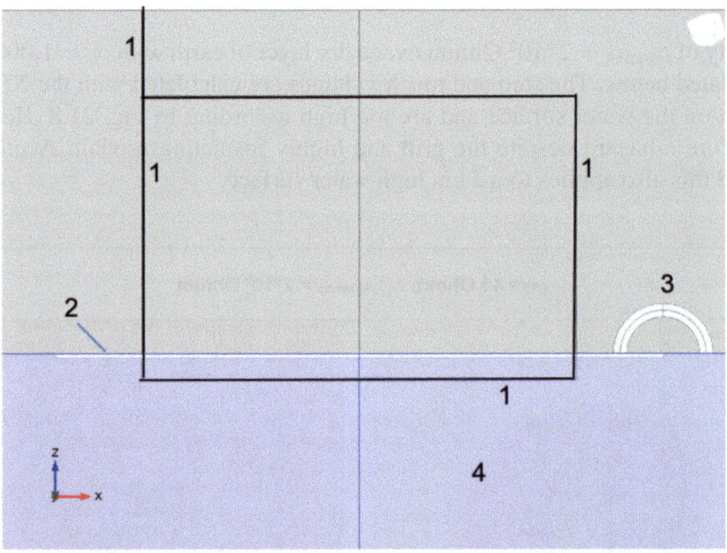

1 Air termination rod, Down conductor und an earthing grid 0,25 m deep
2 Asphalt layer with 5 cm thickness; $\rho_{Asphalt}$ = 250.000 Ohmm
3 Person is represented as a hose with R = 1.000 Ohm
4 Half sphere; ρ_E = 1.000 Ohmm

Fig. 21.9 Cross sectional view of the used 3-D Modell

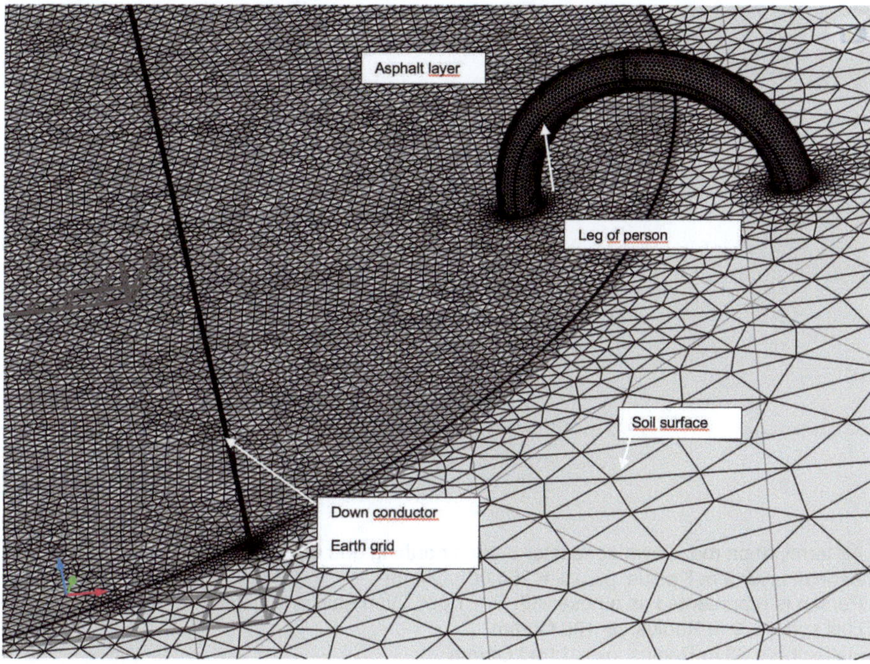

Fig. 21.10 3-D View of the model including meshes

21.9.5 Calculation with Limited and Sprinkled Asphalt Layer with Comsol-Multiphysics

The step voltage in case of heavy rain and limited asphalt area is modelled according to Fig. 21.11. Here a two-layer model with $\rho_E = 1{,}000$ Ohmm below the asphalt surface and with $\rho_E = 100$ Ohmm outside the asphalt surface was used. The person stands on the ground and thus in a 2 or 15 cm high layer of water. The calculation, Table 21.4, shows that in the whole area of the surface the step voltage is too high and that heavy rain is a great danger. The grounding grid and the asphalt layer are rendered ineffective by the well-conducting rain. In a 2-D simulation the grounding grid was simulated as a conductive plate to reduce the computing time. Figure 21.12 shows a strong flow field in the water layer as the reason for the high step tension with water on the asphalt surface.

21.9.6 Technical Solution with Potential Equalisation

The solution with the help of an asphalt layer is only effective with dry asphalt, but is unrealistic. An asphalt layer, covered with rainwater, fails even at a water level

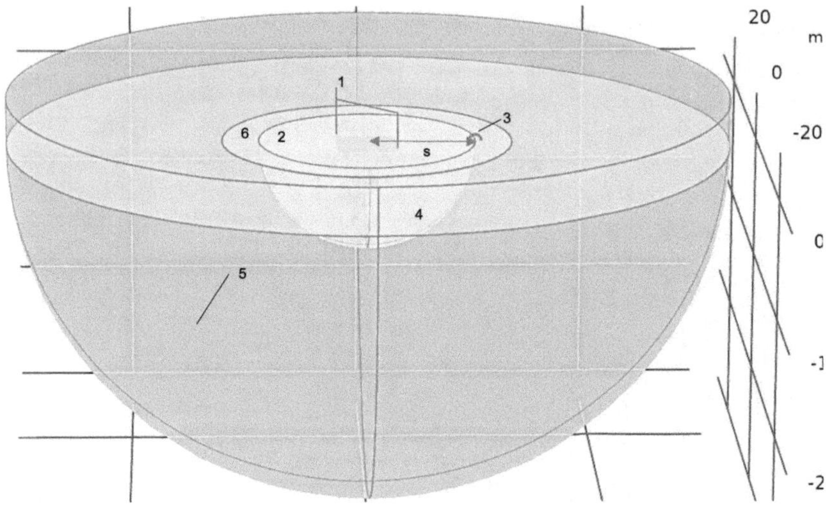

1 Air termination rod, Down conductor und an earthing grid 0,25 m deep
2 Asphalt layer 9 m Radius, $\rho_{Asphalt}$ = 250.000 Ohmm
3 Person is represented as a hose with R = 1.000 Ohm having a distance s from center point
4 Half sphere, 9 m Radius; ρ_E 100 Ohmm
5 Half sphere 30 m Radius; ρ_E = 1.000 Ohmm
6 Water layer 12 m Radius; ρ_W = 45 Ohmm

Fig. 21.11 3-D view of the model for calculating the step voltage with asphalt and water layer

Table 21.4 Step voltage at positions on the earth's surface at distance s according to Fig. 21.11. 100 kA 10/350

Distance s Fig. 21.7 (m)	Step voltage U_S at 2 cm rain water (kV)	Step voltage U_S at 15 cm rain water (kV)
1	76.5	45
2.5	228	159
4.5	272	229
9	112	114.5

of 2 cm. The only reliable way to prevent dangerous step voltages, even above the edges of fine-meshed earth electrodes, is therefore the well-known potential control with the aid of additional earth electrodes.

In the extreme case of a 15 cm high rainwater layer, a safe range for step voltage and partially for touch voltage **can only be achieved** in a two-layer model of the soil with different specific earth resistance and potential control with ring and deep earth electrodes as well as an insulated down conductor according to Fig. 21.13. The insulated down conductor is to be dimensioned for a voltage of approx. 180 kV 10/350. The field picture in Fig. 21.14 was created in a Comsol 2-D simulation and shows the effectiveness of the earth electrodes.

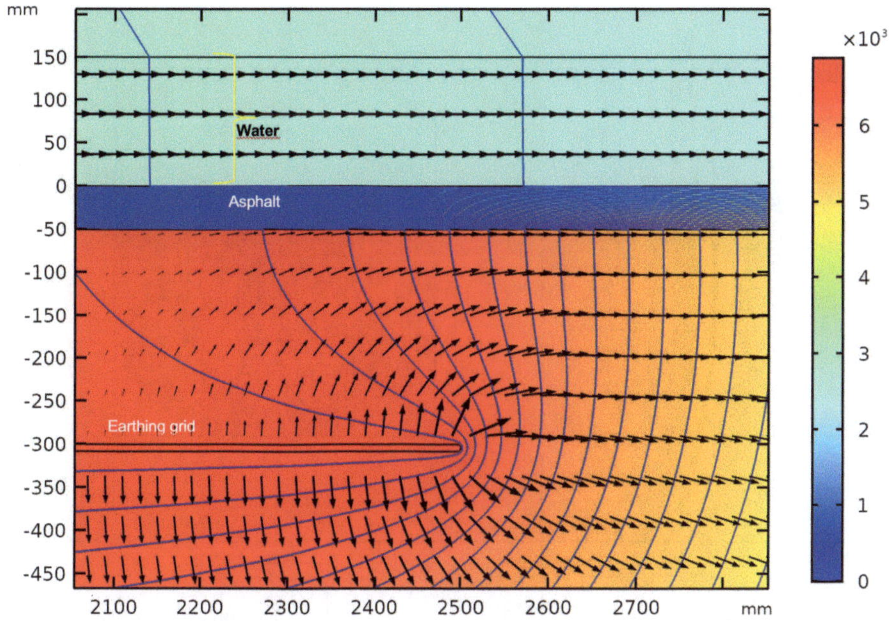

Fig. 21.12 Field plot of the potential, equipotential lines and current density vectors as "arrow-surface" with logarithmic display of the vectors. $\rho_E = 1.000$ Ohmm. Asphalt layer and water layer on top with $\rho_{asphalt} = 2*10^6$ Ohmm; $\varepsilon_{asphalt} = 3$ and rainwater with $\rho_W = 45$ Ohmm; $\varepsilon_W = 80$. Caclulated with 25 kHz

Note: As shown in Fig. 21.12, the lightning current flows into the water surface above the asphalt layer, as this is a good conductor. An insulated down conductor can therefore be used to prevent the current from flowing into the water layer.

21.9.7 Formation of Surface Discharges

With a water-covered asphalt surface, there is a danger to life anyway, so there is no need to assess surface discharges here. On a dry asphalt surface, the formation of surface discharges at the earth entry / discharge is not possible due to the too low field strength of <15 kV/cm.

Above the edge of the earth electrode, a high field strength still occurs on the surface of the asphalt layer just above the field strength for streamer discharges on the earth surface. However, the calculation with the 3-D model has shown that the loaded step voltage just above the edge of the earth electrode is not critical. Sliding discharges are therefore not to be expected in this area and personal safety is ensured.

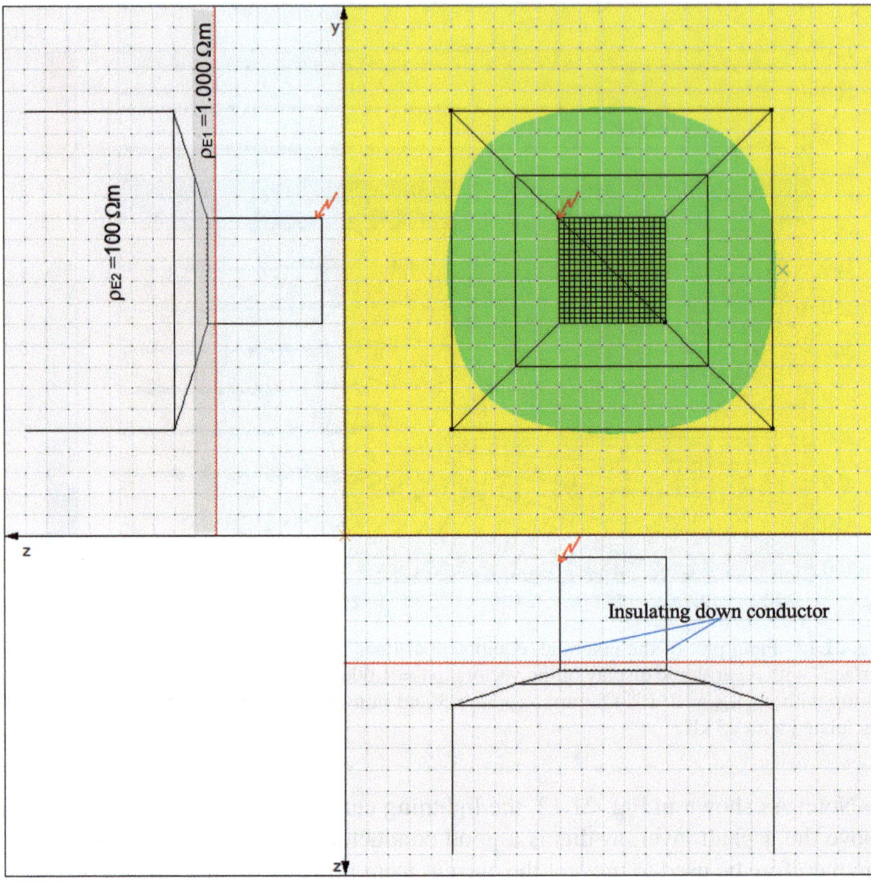

Fig. 21.13 Step and contact voltage for a water layer of 10 cm height on asphalt of 5 cm height with a two-layer model

21.10 Summary

Material	Condition	Step voltage inside shelter	Step voltage outside shelter
None	Dry	OK	
Asphalt layer 5 cm	Dry	OK	OK
	Wet	OK; if: $\rho_{Asphalt}$ > 10.000 Ohmm	
	10 cm rain water		
Layer of gravel 15 cm	Dry	OK	OK
	wet	OK; if: ρ_E < 400 Ohmm, Do not touch down conductor	

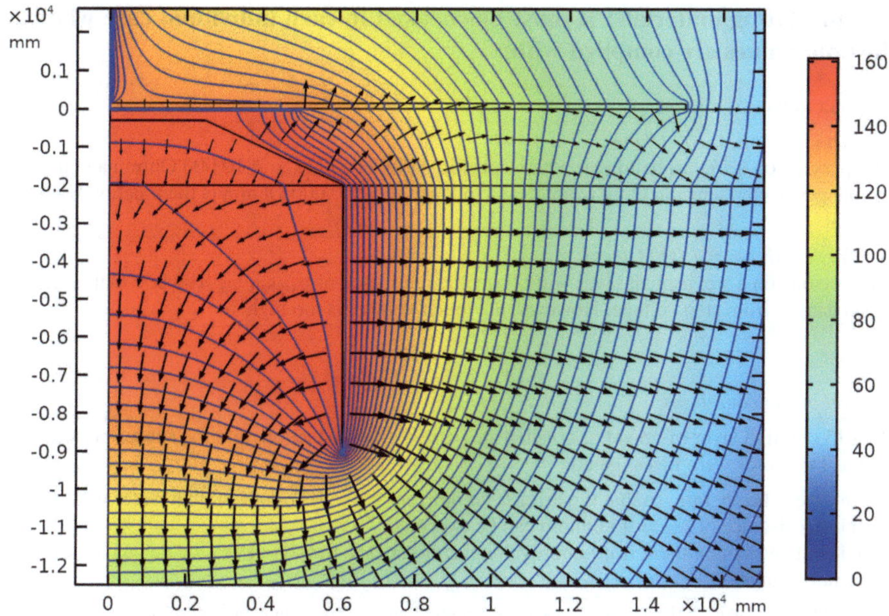

Fig. 21.14 Display of the potential (in colours), equipotential lines and current density vectors as "arrow-surface" with logarithmic display of the vectors for the technical solution as shown in Fig. 21.13. (Calculated with 25 kHz)

Dry insulating materials: The step voltage is controlled within the examined parameters.

Dry asphalt
– *Earth with* ρ_E = *100 Ohmm*: For persons on a dry asphalt layer with a specific resistance of asphalt = 2*10^6 Ohmm and a specific earth resistivity of ρ_E = 100 Ohmm, there is no danger from step and touch voltage
– *Earth with* ρ_E = *1.000 Ohmm*: For persons on a dry asphalt layer with a specific resistance of $\rho_{asphalt}$ = 2*10^6 Ohmm and a specific earth resistance of ρ_E = 1,000 Ohmm, there is no danger from step voltage. In an area outside the earth grid the touch voltage is then too high
Dry gravel
– *Earth with* ρ_E = *100 Ohmm*: For persons on a dry gravel layer with a specific resistance $\rho_{gravel} \geqq$ 100,000 Ohmm and there is no danger of step tension. Outside the earth grid the touch voltage is too high
– *Earth with* ρ_E = *1.000 Ohmm*: Same results as above

Wet insulating materials: It must be checked in each individual case whether the limit values are complied with:

Asphalt in wet conditions[3]
– *Earth with* $\rho_E = 100$ *Ohmm*: For persons on a wet asphalt layer with a specific resistance $\rho_{asphalt} \geqq 10{,}000$ Ohmm and there is no danger of step tension. Only from a value of asphalt >500,000 Ohmm is the touch voltage not critical
– *Earth with* $\rho_E = 1.000$ *Ohmm*: For persons on a wet asphalt layer with a specific resistance of $\rho_{asphalt}$ t $\geqq$ 100.000 Ohmm there is no danger from step voltage. However, the touch voltage is too high

Gravel in wet conditions
In a 15 cm high layer of gravel or crushed stone, water will penetrate into the spaces between them. Two cases were investigated:
Case 1 Very heavy rain, so that the layer is completely soaked through, so that water with a specific resistance of $\rho_W = 45$ Ohmm can be expected The result shows that there is no improvement compared to the soil alone, the dangerous area does not change
Case 2 Rain moistens the layer of gravel or crushed stone. Two materials were investigated:
Material 1 (0.04 m crushed granite with fine fractions) according to the standard IEEE 80-2013 with a specific resistance of $\rho_{S,K} = 1200$ Ohmm at a fraction of rainwater with $\rho_W = 100$ Ohmm The result shows here exclusively with a specific earth resistance of $\rho_E = 50$ Ohmm (this value is rarely reached). Already from $\rho_E \geq 100$ Ohmm on, dangerous areas of step voltage >25 kV are present. The touch voltage is too high in all areas
Material 2 (0.025–0.1 m washed granite similar to gravel) according to the standard IEEE 80–2013 with a specific resistance of $\rho_{S,K} = 5000$ Ohmm at a share of rainwater with $\rho_W = 100$ Ohmm
The result here only shows safe areas with a specific earth resistance of $\rho_E = 200$–300 Ohmm. Dangerous areas of step voltage >25 kV already exist from $\rho_E \geq 300$ Ohmm When calculated with ionisation and $\rho_E = 400$ Ohmm, there is no danger, whereas with $\rho_E = 1{,}000$ Ohmm there is a danger due to too high step voltage. The touch voltage is too high in all areas

Rain water on asphalt layer
The calculation only makes sense for $\rho_E = 1{,}000$ Ohmm, because the soil underneath an asphalt layer dries out
– Rainwater layer with $\rho_W = 45$ Ohmm For persons on an asphalt layer covered with rainwater with a specific resistance of $\rho_{asphalt} = 2*10^6$ Ohmm and a specific earth resistance of $\rho_E = 1{,}000$ Ohmm, there is a danger to life at a water level of 2 cm and >2 cm due to excessive step and touch voltage. Remedy is only possible with a double potential control ring with 4 deep grounding electrodes and insulated lead-in of the derivation/grounding lead-in

[3] It can be assumed that the conductivity is reduced by the water film on the surface.

References

1. DIN EN 62305-3 (VDE 0185-305-3)2011-10
2. Meppelink, J., König, A.: Numerische Berechnung der Berührungs- und Schrittspannung. 10.VDE/ABB-Blitzschutztagung, 2013 in Neu Ulm
3. Meppelink, J., König, A.: Berechnung der Schrittspannung im Blitzrelevanten Frequenzbereich. 11.VDE/ABB-Blitzschutztagung, 2013 in Neu Ulm
4. Deutsche Bauindustrie Bauen und Services sowie bga. Hauptverband der Deutschen Bauindustrie e.V. Bundesfachabteilung Gussasphalt, Kurfürstenstrasse 129, 10785 Berlin. Beratungsstelle für Gussasphalt Anwendung e.V.Dottendorfer Str. 86. 53129 Bonn
5. Oburger, W.: Die Isolierstoffe der Elektrotechnik. Springerverlag 1957
6. IEEE Std 80-2000
7. EN 13242 Gesteinskörnungen für ungebundene und hydraulisch gebundene Gemische für Ingenieur-und Straßenbau
8. ASTM_G57-06(2012-00): Standard Test Method for Field Measurement of Soil Resistivity Using the Wenner Four-Electrode Method
9. DIN EN 62305–1 (VDE 0185-305-1)2011-10
10. Suchanek, Sebastian: Auswirkungen von Schrittspannungen auf den Menschen. 9.VDE/ABB-Blitzschutztagung, 27.-28. Oktober 2011 in Neu Ulm
11. Rock, M.: Grenzwerte für Schritt-und Berührungsspannungen an Blitzschutz-Ableitungseinrichtungen und -Erdungsanlagen. 11.VDE/ABB-Blitzschutztagung, 2015 in Neu Ulm
12. Grcev, L.: Modeling of grounding electrodes under lightning currents. IEEE Trans. EMC **51**(3) (2009)
13. Cigre-Report. Guide to procedures for estimating the lightning performance of transmission lines. WG 01 Study committee 33.01 (1991)
14. Gebhardt, O.: Die Bemessung der Erdung in Hochspannungsanlagen. ETZ-A Bd.96 (1975) H.9
15. VDE/ABB Merkblatt Blitzschutz von Schutzhütten

References

1. DIN EN 50522 (VDE 0101-2):2011-11
2. IEEE Std. 80-2015
3. Rock, M., Zischank, W., Copper, J.: Grenzwerte für Schritt- und Berührungsspannungen an Blitzschutz-Ableitungseinrichtungen und Erdungsanlagen. 11. VDE/ABB-Blitzschutztagung, Neu-Ulm (2015)
4. DIN EN 62305-1 (VDE 0185-305-1)2011-10
5. Kern, A.: Presentation during Conference of VDE/ABB (2020)
6. Hannakam, L.: Berechnung der transienten Stromverteilung im zylindrischen Massivleiter. ETZ-A Bd.91 (1970) H.1 S.50–54
7. www.wolframalpha.com
8. Ruechli, A.E., Antionini, G., Jiang, L.: Skin-effect model for round wires in PEEC. In: International Symposium on Electromagnetic Compatibility EMC Europe (2012)
9. www.Spectrum-soft.com
10. DIN EN 62305-2 (VDE 0185-305-3)2011-10
11. Meppelink, J.: Errichten eines Schutzraumes. Elektropraktiker, Berlin **74**, 2 (2020)
12. DIN EN 62305-3 (VDE 0185-305-3)2011-10
13. XGS-LAB Grounding System Analysis. www.xgslab.com
14. Comsol-Multiphysics, AC-DC Modul www.comsol.de
15. Clayton, R.P.: Inductance Loop and Partial. Wiley (2010)
16. EN TS 62561-8
17. Bischoff, M., Meppelink, J.: Isolierte Blitzschutzsysteme. De 15–16 (2018)
18. Meppelink, J.: Ein Beitrag zur Begrenzung der Schrittspannung durch isolierende Schichten. VDE/ABB-Blitzschutztagung, Neu-Ulm (2015)
19. IEC 60479-5: 2007, Table 1, dc foot to booth hands
20. Yang-Woo, Y. et al.: Soil Discharge characteristics in inhomogeneous field caused by lightning impulse voltages. Journal of the Korean Institute of IIluminating and Electrical Installation Engineers **29**(4), 95–101 (2015)
21. Wiesinger, J.: Funkenstrecken unter Blitz- und Schaltstoßspannungen. ETZ-A. Bd. 90 (1969)
22. Berger, K.: Das Verhalten von Erdungen unter hohen Stoßströmen. Bulletin SEV 37(8) (1946)
23. Nixon, K.J.: The Lightning Transient Behaviour of a Driven Rod Earth Electrode in Multi-Layer Soil. University of the Witwatersrand, Johannesburg
24. Böhme, H.: Mittelspannungstechnik. Huss-Medien GmbH Berlin (2005)
25. Thione, L.: The dielectric strength of large air insulation. In: Ragaller, K. (ed.) Surges in High Voltage Networks. Plenum Press, New York and London
26. Sihvola, A.: Mixing rules with complex dielectric coefficients. Surf. Technol. Appl. **1**, 393–415 (2000)

J. Meppelink, *The Lightning Rod as a Danger*,
https://doi.org/10.1007/978-3-031-30434-7

27. Böhme, H.: Mittelspannungstechnik, Schaltanlagen berechnen und entwerfen. Huss Medien GmbH, Verlag Technik, Berlin. 2. Auflage (2005)
28. Oburger, W.: Die Isolierstoffe der Elektrotechnik. Springer Verlag (1957)
29. www.Comsol.com
30. Meppelink: Comment on TC81 MT8. Comparison of lightning currents simulation methods.pdf
31. WO2020151868 (A1) Blitzstromableitvorrichtung
32. Simony, K.: Theoretische Elektrotechnik. J. Ambrosius Barth (1993)
33. Heinz, C., Piemontesi, M., Salge, G.: Surface discharge along solid dielectrics in atmospheric air. In: Conference Record of the 2000 IEEE International Symposium on Electrical Insulation, Anaheim, CA USA, April 2–5 (2000)
34. Suchaneck, S.: Untersuchungen an Blitzschutzerdungsanlagenunter besonderer Berücksichtigung der Schrittspannung. Dissertation TU Darmstadt (2014)
35. Fußball bei Gewitter ? Richtiges Verhalten im Freien. VDE-Information Blitzschutz. Merkblatt des VDE/ABB
36. Ragaller, K.: Surges in High Voltage Networks. Plenum Press (1980), Seite 200
37. Roth, A.: Hochspannungstechnik, Fünfte Auflage. Springer Verlag (1965)
38. Steinbiegler, H.: Die Stossstromerwärmung unmagnetischer und ferromagnetischer Leiter in Blitzschutzanlagen. Habilitationsschrift, TU München
39. Toepler, M.: Über die physikalischen Grundgesetze der in der Isolatorentechnik auftretenden elektrischen Gleiterscheinungen. Archiv für Elektrotechnik X. Band 5 und 6 Heft (1921)
40. Shimazaki, T.: Itaru Tsuneyasu, flashover mechanism on the surface of a dielectric plate in the atmosphere under negative impulse voltage. Electr. Eng. Jpn. **105**(4) (1985)
41. Rosenlöcher, P.: Untersuchung von Oberflächenentladung bei Stoßspannung. Mitteilungen aus dem Elektrotechnischen Institut der Technischen Hochschule Aachen (1931)
42. Dashuk, P.N., Emel´yanov, S., Ivanova, T.A.: Creepage discharge along the surface of solid dielectrics in water. Soviet Physics Journal. Izvestiya VUZ. Fizika **11**(2), 111–117 (1968)
43. Auerbach, P.S.: Wilderness Medicine, 7th edn. Elsevier (2016). ISBN-10-0323359426
44. Cooper, M.A., Holle, R.L.: Mechanisms of lightning injury should affect lightning safety messages. In: 2010 21st International Lightning Detection conference. Orlando, Florida, USA (2010)
45. Blumenthal, R.: Injuries and deaths from lightning. Department of forensic medicine, University of Pretoria
46. Cooper, M.A., Holle, R.L.: Reducing lightning injuries world wide. Springer Verlag (2019)
47. Chowduri, P., Mishra, A.K., Mc. Connell, B.W.: Volt-time characteristics of short air gaps under nonstandard lightning voltage waves. IEEE Trans. Power Deliv. **12**(1) (1997)
48. Wiesinger, J.: Absinken der Festigkeit von Funkenstrecken bei steilen Stoßspannungen. Bull. ASE **58**(1967)3, 4 février
49. Bibliography of research on nonstandard lightning voltage waves. IEEE Trans. Power Deliv. **9**(4) (1994)
50. Review of research on nonstandard lightning voltage waves. Task force 15.09. IEEE Trans. Power Deliv. **9**(4) (1994)
51. Zischank, W., Wiesinger, J., Hasse, P.: Insulators for isolated or partly isolated lightning protection systems to verify safety distances. 23. ICLP, Florenz, S. 513–518 (1996)
52. Neuhaus, H. Pigler, F.: Blitzkennwerte als Grundlage der Bemessung von Blitzschutzmßnahmen. Etz Bd.103 (1982) Heft 9
53. Meppelink, J.: A review of the determination of the safety distance for lightning protection systems using the area time law. In: 29th International Conference on Lightning Protection 23rd–26th. Uppsala, Sweden (2008)
54. IEC 60060-1
55. Guide to procedures for estimating the lightning performance of transmission lines. Cigre Report 63
56. Mishra, C.S.: Die Stoßspannungsentladung bei Beregnung. Dissertation TU München (1962)
57. Berger, K.: Comparative impulse test with impulse voltage on spark gaps. Cigre Bericht Nr. 326 (1956)

58. Heidler, F., Zischank, W.: Necessary separation distance for lightn ing protection systems—IEC 62305–3 revisited. X SIPDA Curitiba, Brazil (2009)
59. Ushakov, V.Y.: Insulation of High-Voltage Equipment. Springer-Verlag, Berlin and Heidelberg GmbH & Co. KG (2010). ASIN: B017WOLFJ6
60. Andrews, C.J., Darveniza, M.: Telephone mediated lightning injury: an Australian survey. J. Trauma 29(5):665–667
61. Source: Die Elektrizität des Himmels und der Erde. Dr. Alfred Ritter von Urbnitzki, A. Hartleben´s Verlag (1888)

MIX
Papier aus verantwortungsvollen Quellen
Paper from responsible sources
FSC® C105338

If you have any concerns about our products,
you can contact us on
ProductSafety@springernature.com

In case Publisher is established outside the EU,
the EU authorized representative is:
Springer Nature Customer Service Center GmbH
Europaplatz 3, 69115 Heidelberg, Germany

Printed by Libri Plureos GmbH
in Hamburg, Germany